INTERDISCIPLINARY MATHEMATICS

VOLUME XVI

QUANTUM AND FERMION DIFFERENTIAL GEOMETRY

PART A

ROBERT HERMANN

MATH SCI PRESS
53 JORDAN ROAD
BROOKLINE, MA. 02146

Copyright © 1977 by Robert Hermann
All rights reserved

ISBN 0-915692-22-8

MATH SCI PRESS
53 JORDAN ROAD
BROOKLINE, MA. 02146

Printed in the United States of America

PREFACE

Some of the most exciting and fruitful periods in the history of science occur when new mathematical methods are successfully introduced into the study of fundamental physical phenomena. To illustrate, think of the great "supernovae" of mathematical physics--Newtonian mechanics, relativity, and quantum mechanics.

The theory of "elementary particles" has been under intensive development for the last thirty years. Despite the strong probability (through historical evidence) that new mathematical paradigms would be needed to achieve a fundamental understanding, physicists have been very conservative (and "pragmatic") about the sort of mathematics they consider admissable. To my eye, it has been evident since the introduction of sophisticated Lie group theory in the early 1960's that fundamental new mathematics, particularly concerning Lie groups and differential geometry, would be required. In the last five years there have been incremental development by physicists of a succession of ideas (gauge-Yang-Mills fields, solitons, instantons, "super" spaces,...) that I believe add up to a new way of applying geometry to physics, and that give some clues as to what a *mathematically* unified physical theory will look like.

If it is so self-evident that the answer to the fundamental problems of physics will come from geometry, why haven't physicists learned it? I am afraid that a good deal of the answer must be ascribed to the perversity of the physicists' dominant "pragmatic" mathematical ideology. Pragmatists notoriously have no esthetic sense, and "modern" differential geometry (as opposed, say, to tensor analysis, which it replaced and which physicists do know) requires, above all, some aesthetic feeling for the structure of modern mathematics. It does not help either that (at least in the U.S.) the formal mathematical education of most physicists stops at the advanced calculus level. Of course, another perverse feature has been the complete inversion and isolation of the small community of differential geometers. Their fathers and grandfathers took great delight and inspiration in mixing it up with the physicists, but the current generation has completely lost that taste and talent.

The mathematicians who have been most interested in "fundamental physics" have been trained in functional analysis, which is, in my view, not necessarily the most creative source of mathematical ideas for physics. These mathematicians often think their highest goal is to add "mathematical rigor" to physics. However, this is usually <u>not</u> the goal of geometers--good conceptual ideas usually comes first and the "rigor" can be put in later. This distinction goes back to Poincaré; he wrote of two types of mathematician--the "analyste" and the "géomètre". He did not hide his opinion of which type he thought was more creative! (Keep in mind that this does not excuse sloppy mathematical thinking, notation, formulation, etc. In fact, for the "géomètre" these are usually more important than for the "analyste".) Another characteristic of the "modern" geometric point of view is a strong interplay between *algebra* and *geometry*. Ideally, someone wanting to apply modern geometry should have at least a smattering of knowledge of all three main "technical" branches of modern mathematics-- analysis, algebra, and topology. This is, of course, a great barrier for the physicist or engineer, who at best will have only a strongly focussed and

selected knowledge of mathematics. (Many physicists have told me that they
find these geometric matters very interesting, but they do not feel that they
can afford to invest the time to acquire the varied mathematical background
they would need.)

One of the prime advantages of thinking about science in this "geometric"
way is the overview one obtains of the essential *unity* of science *and* technology.
The reader will notice that I intersperse in this series books on physics with
books on systems theory. In fact, they have a common mathematical (and methodological) core. Ironically, it is the *engineer* who has pursued most forcefully
the general setting of the ideas--physics (i.e., elementary particle physics)
has become the intellectual slum of science because physicists have given up
the attempt at a broad, long-range view and devote most of their energies to
pursuit of one fad after another. I see more effort going into public relations and advertising (e.g., cutesy-wutesy names) than anything else.

However, the reader will notice that I am obsessed with physics myself.
It is still *the* subject which offers the most to the mathematician. Just when
I decide to give up the whole business and devote myself wholeheartedly to
systems theory and nonlinear waves (subjects with honest engineering ancestry
and much less pretension and phoniness) something interesting pops up! Recently,
this has involved two subjects--the theory of "superspace" (which I prefer
to call "fermion" differential geometry) and the "Moyal" approach to quantization. In fact, the theme of this volume is that they have a mathematical
unity--as a sort of "generalized" differential geometry, where one sets aside
geometry in terms of "points" and creates an *algebraic* structure which includes
"ordinary" manifolds as special cases, but allows for other possibilities.

This potential for generalization has, of course, always been evident in
"modern" differential geometry. We differential geometers have been diffi-
dent and modest about this--but there have been various wild men (mainly
algebraic geometers and differential topologists) who have made their fame
and fortune with variations on this theme. However, it seems to me that none
of these attempts are as *interesting* as the directions suggested by *physics*--
again we see the principle that it is the sophisticated and complex feedback
interconnection between science and mathematics that can give *direction* to
mathematics, and, at the same time, provide new mathematical tools for use in
science.

Since this volume contains rather recent research, it has been done in
a different way. It is basically a collection of papers I wrote in the
academic year 1976-1977. (The first chapter is earlier work.) I originally
wrote them for journals, then decided they had a methodological unity and that
it would be a pity to break them up. The underlying research work was supported
in part by the National Science Foundation, Grant MCS-07993. I was a vistor in
the Harvard University Physics Department. I would like to thank Professor
S. Glashow for his hospitality. I am grateful to N. Wallach for conversations
about the Moyal product. Again, I thank Karin Young for her fine typing!

TABLE OF CONTENTS

Page

PREFACE iii

I. THE POISSON-MOYAL BRACKET, CLASSICAL INVARIANT THEORY AND THE DEFORMATION OF QUANTUM INTO CLASSICAL MECHANICS 1

1. Introduction 3
2. The "Überschiebung" Operator of Classical Invariant Theory 6
3. The Poisson-Moyal Bracket and the Weyl Correspondence Between Classical and Quantum Observables 10
4. Some Remarks about Quantization on a General Symplectic Manifold 13

Bibliography 16

II. QUANTUM AND FERMION DIFFERENTIAL GEOMETRY A LA CARTAN 19

1. Introduction 21
2. Differential Forms for an Associative Algebra 22
3. Boson Quantum Differential Forms 24
4. Hamiltonian Particle Mechanics 25
5. Lagrangian Field Theories 29

Bibliography 31

III. GENERALIZED BEREZIN-KAC GROUPS IN QUANTUM DIFFERENTIAL GEOMETRY 33

1. Introduction 35
2. Generalizations of Lie Groups Based on Associative Algebras 35
3. Boson Quantum Lie Groups 37
4. Generalized Homogeneous Spaces 40

Bibliography 41

		Page
IV.	THE DIFFERENTIAL-GEOMETRIC FOUNDATIONS OF "SUPERSPACE" AND "SUPERSYMMETRY" THEORIES, I. GRADED LIE ALGEBRAS AND THEIR GEOMETRIC REPRESENTATIONS	43
	1. Introduction	45
	2. Lie Algebras of the Simplest "Current Algebra" Type, with no Schwinger Terms	46
	3. The Tensor Product of a Grassman Algebra and a Lie Algebra. The Definition of a Graded Lie Algebra	48
	4. Linear Representations of a Graded Lie Algebra	53
	5. Derivations of the Exterior Algebra of a Vector Space	55
	6. Graded Lie Algebras as "Vector Fields" in Fermion Spaces	59
	7. The Beltrami Bracket on a Riemannian Manifold	61
	8. Z_2-Graded Lie Algebras	64
	9. Z_2-Graded Lie Algebras which Act on Riemannian-Symplectic Manifolds	66
	10. Kähler Manifolds	68
	11. The Hermitian-Skew Hermitian Z_2-Graded Lie Algebra Realized on Projective Space	70
	12. Graded Lie Algebras for Classical Mechanical Problems	72
	13. Linear Heisenberg-Fermi Structures on a Vector Space and their Fock Representations	78
	Bibliography	84
V.	THE DIFFERENTIAL-GEOMETRIC FOUNDATIONS OF "SUPERSPACE" AND "SUPERSYMMETRY" THEORIES, II. THE QUADRATIC DIFFERENTIAL CO-FORMS ON THE DUAL SPACE OF A GRADED LIE ALGEBRA	87
	1. Introduction	89
	2. Co-Riemannian and Co-Symplectic Structure. Quadratic Differential Co-Forms	89
	3. Morphisms and Products of Quadratic Differential Co-Forms	92
	4. The Dynamics Determined by a Quadratic Differential Co-Form and a Function	93
	5. The Quadratic Differential Co-Forms on the Dual Space to a Graded Lie Algebra	94
	6. Tensorial Functions on the Dual Space to a Graded Lie Algebra	97
	7. The Equations of Motion Determined by a Quadratic Tensor	98
	Bibliography	101

Page

VI. THE DIFFERENTIAL-GEOMETRIC FOUNDATIONS OF
"SUPERSYMMETRY" AND "SUPERSPACE" THEORIES, III.
LAGRANGIAN FIELD THEORIES IN FERMION AND QUANTUM
DIFFERENTIAL GEOMETRY   103

1. Introduction   105
2. Differential Form Algebras   107
3. Boson and Fermion Differential Form Algebras: "Superspace"   108
4. General Heuristic Principles about the Construction of Differential Form Algebras   111
5. The Cartan Differential Forms and Equations of Motion for Variational Problems in "Boson" Differential Geometry   114
6. Lagrangian Field Theory for Fermion Fields   117
7. Linear Equations   120
8. Lie Derivative and Contraction for Differential Form Algebras   122
9. One-Variable Calculus of Variations, Cauchy Characteristic, Hamilton-Jacobi Equation, Hamilton Equation, Poisson Bracket, etc.   129
10. The Dirac Equation and Poisson Bracket of Fermion Spaces   135
    Bibliography   139

VII. DIFFERENTIAL FORMS FOR CERTAIN ASSOCIATIVE ALGEBRAS   143

1. Introduction   143
2. Differential Forms on a Manifold   144
3. Generalities about Graded Algebras   147
4. Differential Form Algebras   149
5. Commutativity Properties   150
6. The Boson Differential Form Algebra   153
7. The Fermion Differential Form Algebra   154
8. A Change in Notation with $\mathscr{A}$ Replacing $\mathscr{A}^0$, $\mathscr{D}^r(\mathscr{A})$ Replacing $\mathscr{A}^r$ 156
9. The Z_2-Graded Structure for the Boson Differential Form Algebra   157
10. The Commutation Relations and Parity Operator for Grassmanian Differential Form Algebras   160

VIII. THE POISSON-MOYAL BRACKET AND QUANTUM SYMPLECTIC GEOMETRY   165

1. Introduction   165
2. The Transvection Operations   167

	Page
3. Some Traditional Special Cases of the Transvection Operation	168
4. Covariance Properties of the Transvection	171
5. The Associative Moyal Product and the Poisson-Moyal Lie Algebra	175
6. The Moyal Bracket Stands to "Quantum Symplectic Geometry" as the Poisson Bracket Stands to "Ordinary Symplectic Geometry"	178
7. A New Definition of "Quantum Dynamics"	183
Bibliography	184

IX. GENERALIZATIONS OF THE POISSON-MOYAL BRACKET. DIFFERENT METHODS OF QUANTIZATION   185

1. Introduction   185
2. Generalized Transvection Operations and Moyal Products   185
3. The Point of View of Deformation Theory   186
4. The Moyal Structure on the Dual Space of a Lie Algebra   192

I.

THE POISSON-MOYAL BRACKET, CLASSICAL INVARIANT THEORY AND THE DEFORMATION OF QUANTUM INTO CLASSICAL MECHANICS.

ABSTRACT

The Weyl quantization procedure--related to the definition of a Poisson-Moyal Bracket operator--is interpreted in a differential geometric manner, leading to a new differential-geometric structure which will be called a <u>higher order symplectic structure</u>. In the case of the usual linear symplectic manifold--and requiring invariance under linear symplectic automorphisms--the Poisson-Moyal bracket is defined by invariant differential operators which have already appeared in 19th Century Invariant Theory. The geometric structures involved can be generalized to arbitrary manifolds. This suggests a mathematical scheme for defining quantum mechanics in a purely differential geometric, non-Hilbert space manner, reducing to classical mechanics as $\hbar \to 0$ in the sense of deformation theory.

1. Introduction

Consider a classical mechanical system with n degrees of freedom. Let $q = (q_1, \ldots, q_n)$ be its configuration coordinates, $p = (p_1, \ldots, p_n)$ its momentum coordinates. Then (p,q) denotes a point of R^{2n}. Consider C^∞, real-valued functions $f, f' : R^{2n} \to R$. Denote them by $F(R^{2n})$.

The classical <u>Poisson bracket</u>, defined as usual,

$$(f, f') \to \frac{\partial f}{\partial p} \frac{\partial f'}{\partial q} - \frac{\partial f}{\partial q} \frac{\partial f'}{\partial p} \equiv \{f, f'\}$$

defines a Lie algebra structure on $F(R^{2n})$. The role that this Lie algebra plays in classical mechanics has been well-known for over one hundred years.

Since the first days of quantum mechanics it has been apparent that the representations by Hermitian operators of this Lie algebra or certain Lie subalgebras has something to do with what one might call the "process of quantization of a classical mechanical system". However, the precise relation has remained obscure; one knows how to deal satisfactorily with a few systems which, luckily, turn out to be the most important ones in practice, but nothing about the general situation. Of course, this is typical of the physicist's "pragmatic" attitude toward mathematical problems. (Incidentally, this "pragmatism" has the unfortunate consequence of leaving Physics stranded in its own mathematical ghetto while the rest of the pure and applied mathematical world develops independently. This isolation has unfortunate implications for progress in fundamental physics, especially in elementary particle physics. Judged on a historical scale, there have been no real advances since the understanding of renormalization of quantum electrodynamics; everything since then has been more or less sophisticated (and expensive!) phenomonology. Physicists have, of course, been as inventive as ever in terms of new ideas, but without serious and long term mathematical follow-through they go nowhere, or last as long as the next experiment and change in intellectual fashion.) The basic work on this quantization problem in the post-World War II period has been a memoire by L. Van Hove[1] that is very obscure, hence has been almost ignored in the recent literature. He showed that it was not possible to represent <u>all</u> of $F(R^{2n})$ in a way that followed the usual physical rules. Using more-or-less classical ideas involving "contact" and "canonical" transformations he did construct a representation of the entire Poisson bracket algebra but noticed that it was not satisfactory for physics. More recently, Van Hove's representation has been revived--in various global versions--by many authors, particularly by Segal,[2] Konstant,[3] and Souriau,[4] usually without mention of the representation for physical purposes. Prosser[5] has presented an example which emphasizes this point. In various cases, the representation constructed in this way can be modified to give the usual, physically satisfactory, way of quantization, but this does very little to solve the general problems. The basic question can be stated as follows: Is there some way of defining an interesting category of classical mechanical system with "quantization" appearing as a functor to a suitable "category" of "quantum" systems?

In this work I plan to evade this long-standing problem and strike off in a different direction, suggested by the work of Weyl[6] and Moyal.[7] In differential geometric language what they do can be described in the following way.

Replace the Poisson bracket with a one-parameter family of Lie algebra structures on $F(R^{2n})$, which depends on Planck's constant $2\pi\hbar$ and reduces to the Poisson bracket for $\hbar = 0$. It can be satisfactorily represented by Hermitean operators on a Hilbert space, and has the following form:

$$\{f,g\} = \{f,g\} + \sum_{m=1}^{\infty} D_{2m+1}(f,g)\hbar^{2m} \qquad (1.2)$$

where D_j is a j-th order bilinear differential operator.

$$F(R^{2n}) \times F(R^{2n}) \to F(R^{2n}) \qquad (1.3)$$

We shall call this the <u>Poisson-Moyal bracket</u>. Because of its genesis in terms of the Weyl correspondence between $F(R^{2n})$ and Hermitean operators, <u>it must be invariant under linear symplectic automorphisms of</u> R^{2n}, hence, so must each of the differential operators D_j which appear on the right hand side of (1.2). Now, 19th century theory invariant theory dealt with such invariant differential operators. The main point of this paper is that these operators already appeared in the 19-th

century work--they were called "Überschiebung" operations.
(In the English literature they were called "transvections".)
This interpretation suggests very general approaches and insights
into the quantization problem, which are to be presented in this
series of papers.

I am indebted to M. Ackerman, A. Mayer, A. Voros, and N. Wall
for conversations about some of these issues.

2. The "Überschiebung" Operator of Classical Invariant Theory

It is not widely appreciated or understood, but "classical
invariant theory" may be interpreted as the study of the action
of linear transformation groups on linear or multilinear differential operators. Thus, in spirit, it has much in common with
modern work on symmetries of particles and fields, classical or
quantum.

Sophus Lie was the leader in interpreting invariant theory
in this way. (In fact, he extended the ideas to cover non-linear
transformation groups and differential equations.) His exposition
(in Ref. 8, Chapter 23) is one of the clearest and most under-
standable for the modern reader of any with which I am familiar.
In particular, we will use his formula for the "Überschiebung"
operator as a differential operator, which I have found nowhere
else. (It is usually given in the "symbolic" notation of
Aronhold, which is difficult for the modern reader to understand.)

Let us begin with the general situation, using standard
differentiable manifold ideas and notation. Let M be a manifold,

with G a transformation group acting via diffeomorphisms on M. Let F(M) denote the C^∞, real-valued functions on M. A <u>differential operator</u> is a mapping D: F(M) → F(M) which is given in local coordinates in the usual way, in terms of partial derivatives. (For linear operators a coordinate-free definition is given in Ref. 10.)

Denote the transform $p \in M$ by $g \in G$ as gp. Let g act on F(M) as follows:

$$g(f)(p) = f(g^{-1}p)$$

A differential operator D: F(M) → F(M) is said to be <u>invariant under</u> G if

$$D(g(f)) = gD(f) \tag{2.1}$$

for $g \in G$, $f \in F(M)$.

(In the classical literature this is usually called a "covariant". Another word used in the modern mathematical literature is "Intertwining Map".)

One can now define various types of "non-linear" operators using ideas of multilinear algebra. For example, we will deal with "bilinear" operators, since it is those which appear in the Poisson-Moyal bracket.

<u>Definition</u>. A map

$$D: F(M) \times F(M) \to F(M)$$

is said to be a <u>bilinear</u> G-<u>invariant differential operator</u> if the following conditions are satisfied:

   a) D is an R-bilinear map, in the sense of linear algebra
   b) D is a differential operator in the sense that in local
      coordinates for $f_1, f_2 \in F(M)$, $D(f_1, f_2)$ depends on the
      finite number of partial derivatives of f_1 and f_2.
   c) $$gD(f_1, f_2) = D(gf_2, g(f_2)) \qquad (2.2)$$
      for $g \in G$; $f_1, f_2 \in F(M)$

To construct the examples of this type of operator, let

$$M = R^2$$

(p,q) be coordinates for M

For each value of $m = 1, 2, \ldots,$ set:

$$D_m(f_1, f_2) = \frac{\partial^m f_1}{\partial p^m} \frac{\partial^m f_2}{\partial q^m} - \binom{m}{1} \frac{\partial^m f_1}{\partial p^{m-1} \partial q} \frac{\partial^m f_2}{\partial p \partial q^{m-1}}$$

$$+ \cdots \pm \frac{\partial^m f_1}{\partial q^m} \frac{\partial^m f_2}{\partial p^m} \qquad (2.3)$$

D_m is called the m-th "Überschiebung" operator. Notice that it reduces to Poisson bracket for $m = 1$. We shall see in the next section that (up to a constant factor) the odd order operators are the ones that appear on the right hand side of the Poisson-Moyal bracket.[1,2] The even order ones appear in physics in the quantization of Fermions. (This does not seem to be in the literature, to my knowledge, and will be considered in a later paper of this series.)

Formula (2.4) is especially interesting from the point of view of classical invariant theory, for $m = 2$. For $f_1 = f = f_2$

$$\frac{1}{2} D_2(f,f) = \frac{\partial^2 f}{\partial p^2} \frac{\partial^2 f}{\partial q^2} - \left(\frac{\partial f}{\partial p \partial q}\right)^2$$

$$= \text{the Hessian of } f.$$

This is, of course, an example of a non-linear invariant differential and it played an important role in the 19th century, particularly in the geometric applications of Invariant Theory.

Let G be the group consisting of translations in R^2, composed with linear maps $R^2 \to R^2$ of determinant one. With regard to the usual symplectic manifold structure, $dp \wedge dq$, M is a symplectic homogeneous space. Of course, D_1 is invariant under G --it is the Poisson bracket. It is now readily seen that D_m, for each m, is also invariant.

The generalization of formula (2.3) for systems of n degrees of freedom should be obvious. $p = (p_1,\ldots,p_n)$, $q = (q_1,\ldots,q_n)$ are interpreted vectorially. The resulting D_m is seen to be invariant under the linear symplectic group

$$Sp(n,R) \ .$$

It is also obviously invariant under translations, hence is

invariant under the group G which is a semi-direct product of $Sp(n,R)$ and the group R^{2n} of translations. This, of course, again makes R^{2n} into a symplectic homogeneous space. Other symplectic homogeneous spaces appear in physics--their invariant bilinear differential operators will be discussed in later papers of this series. Conceivably, one might extend the material to infinite dimensional symplectic manifolds, thus developing methods to deal with quantum field theory.

3. <u>The Poisson-Moyal Bracket and the Weyl Correspondence Between Classical and Quantum Observables</u>

For the sake of completeness I will now sketch the well-known material on the generation of the Moyal relations by means of the Weyl correspondence. For simplicity of notation, we consider only systems with one degree of freedom: Again, the methods generalize to n degrees of freedom using vectorial notation. (p,q) then denote coordinates of R^2, the classical "state" or "phase" space. Let (x,y) denote the dual (in the sense of Fourier transform) linear coordinates of R^2. Let

$$RD(R^2)$$

denote the space of rapidly decreasing, real-valued C^∞ functions on R^2. Thus, $f \in RD(R^2)$ can be written as follows:

$$f(p,q) = \int_{R^2} \hat{f}(x,y) e^{i(xq+yp)} \, dx \, dy \qquad (3.1)$$

$\hat{f}$ is the <u>Fourier transform</u> of f. It is known to be a rapidly decreasing complex-valued C^∞ function on R^2 such that:

$$\hat{f}(x,y)^* = \hat{f}(-x,-y) \tag{3.2}$$

Now, let P,Q be fixed (once Planck's constant is fixed) self-adjoint operators on a Hilbert space $\underset{\sim}{H}$ which act irreducibly and satisfy the Heisenberg commutation relations:

$$[P,Q] = i\hbar \tag{3.3}$$

Associate to $f \in RD(R^2)$ the following self-adjoint operator:

$$\underset{\sim}{f}_\hbar = \int_{R^2} \hat{f}(x,y) e^{i(xQ+yP)} \, dx \, dy \tag{3.4}$$

The Poisson-Moyal bracket is now defined as that operation

$$(f_1 f_2) \to \{f_1, f_2\}_\hbar$$

on functions which makes the mapping

$$f \to \underset{\sim}{f}_\hbar$$

of functions into operators into a Lie algebra homomorphism. In other words, we should have

$$\{f_1, f_2\}_\hbar = \frac{1}{i\hbar} [\underset{\sim}{f}_{1\hbar}, \underset{\sim}{f}_{2\hbar}] \tag{3.5}$$

We shall verify that the terms in the expansion (1.2) for the Poisson-Moyal bracket involve the differential operators (2.3) in odd orders.

Recall first the following well known relation:
$$e^A e^B = e^{A+B+\frac{1}{2}[A,B]} \tag{3.6}$$
which holds for operators A, B such that
$$[[A,B],A] = 0 = [[A,B],B] \tag{3.7}$$
We can calculate the commutator using (3.3), (3.6) (and the fact that the functions are rapidly decreasing which enables limits to be interchanged freely) to calculate the basic Moyal formula:
$$[\underset{\sim}{f}_\hbar, \underset{\sim}{f}'_\hbar] = \int \hat{f}(x,y) \hat{f}'(x',y') e^{i((x+x')Q+(y+y')P)} \\ \frac{2}{\hbar} \sin \frac{\hbar}{2}(yx' - xiy) \, dx \, dy \, dx' \, dy' \tag{3.8}$$

Now let us compute the operator assigned to $D_m(f,f')$, as given by (2.3):

$$\frac{\partial^m f}{\partial p^{m-j} \partial q^j} = (i)^m \int \hat{f}(x,y) y^{m-j} x^j e^{i(xq+yp)} \, dx \, dy$$

$$\frac{\partial^m f'}{\partial q^{m-j} \partial p^j} = (i)^m \int \hat{f}'(x',y') x'^{m-j} y'^j e^{i(x'q+y'p)} \, dx' \, dy'$$

Thus,
$$D_m(f,f') = (-1)^m \int \hat{f}(x,y) \hat{f}'(x',y') (yx' - y'x)^m \\ e^{i((x+x')q+(y+y')p)} \, dx \, dy \, dx' \, dy' \tag{3.9}$$

According to the definition, the operator assigned to $D_m(f,f')$ by the Weyl rule is obtained from (3.9) by making the substitutions

$$p \to P$$

$$q \to Q \; .$$

Comparing to (3.8), we see that operators D_{2m+1} on the right hand side of (1.2) are indeed (up to a constant) the "Überschiebung" operators.

4. Some Remarks about Quantization on a General Symplectic Manifold

What we have done basically is to reinterpret the Weyl quantization procedure so that it makes sense on a general symplectic manifold. It is well known that the Poisson bracket itself can be defined on a symplectic manifold--this is the basic reason why symplectic manifolds are important for physics--but I know of no general geometric structure which will provide operators like D_{2m+1}, which can be used in formula (1.2) to define a Lie algebra structure on $F(M)$. In fact, what is involved is some geometric structure for M which is a "higher order" symplectic structure. I will come back to this in a later paper. (They do not seem to have been discussed before in the literature.)

Having been given differential operators D_{2m+1} on a general symplectic manifold M, and defining a Poisson-Moyal bracket-like operation by means of (1.2), a basic requirement is of course, that it forms a <u>Lie algebra</u>, i.e., satisfies the Jacobi identity. The conditions for this are familiar in the mathematical literature of the theory of "deformation". See Refs. 12 and 13 for treatment of certain parts of it which are especially relevant for physics. The key qualitative fact to keep in mind is that calling $\underset{\sim}{L}$ the Lie algebra formed by F(M), with Poisson bracket as the Lie bracket, the maps

$$D_{2m+1}: \underset{\sim}{L} \times \underset{\sim}{L} \to \underset{\sim}{L}$$

satisfy--when interpreted as "two-cochains of $\underset{\sim}{L}$ with coefficient on the adjoint representation of $\underset{\sim}{L}$"--certain Lie algebra cohomology relations. Another interesting fact is that this sort of "cohomology of infinite dimensional Lie algebras determined by differential operators" occurs in the differential geometric theory of Current Algebras and Gauge Fields. See Refs. 14, 15, and 16. Bloore, Assimakopoulos and Ghobrial have recently also related[17] the quantization problem to deformation theory.

If this program--which I hope to pursue in detail in later papers in this series--succeeds, it might have the physical consequence of pointing out that "quantization" is not a well-defined (in the functorial sense) procedure. I would like to ask what such a possibility would say about the conceptual foundations of quantum mechanics. In certain cases, e.g., where M is a coset space G/H, and G acts as automorphism of the generalized Poisson-Moyal brackets, one might hope to find "uniqueness" of the quantization procedure.

However, systems with an infinite number of degrees of freedom might be particularly prone to non-uniqueness. Certainly recent experimental discoveries in Elementary Particle Physics gives one ample justification for the belief that all of the mysteries of the "quantization" procedure have not been understood by physicists and that there is much meaningful mathematical work to be done!

Bibliography

1. L. Van Hove, Sur certaines réprésentations unitaires d'un group infini des transformations, Acad. Roy. Belg. Cl. Sci. Mem., No. 6, ;951.

2. I. Segal, Mathematical problems of relativistic physics, American Math. Society, Providence, R.I., 1963.

3. B. Kostant, Quantization and unitary representations, Lectures in Mathematics, Vol. 170, Springer-Verlag, Heidelberg, 1970.

4. J. Souriau, Structure des systèmes dynamiques, Dunod, Paris, 1970.

5. R. Prosser, J. Math. Phys., $\underline{5}$, 701 (1964).

6. H. Weyl, The theory of groups and quantum mechanics, Methuen, London, 1931.

7. J. Moyal, Proc. Camb. Phil. Soc., $\underline{45}$ (1949), 99-124.

8. S. Lie, Vorlesungen über continuerliche Gruppen, Chelsea N.Y., 1969.

9. R. Hermann, Differential geometry and the calculus of variations, Academic Press, 1963 (Out of print. A new edition is in preparation, to be published by Math Sci Press, Brookline, Mass.)

10. R. Hermann, Geometry, physics and systems, M. Dekker, N.Y., 1973.

11. R. Hermann, J. Math. Phys., 13 (1972), 833.

12. R. Hermann, Lie groups for physicists, W.A. Benjamin, Reading, Mass., 1966.

13. R. Hermann, Comm. Math. Phys., 2 (1966), 251; 3 (1966), 53; 3 (1966) 75; 5 (1967), 131; 5 (1967) 157; 6 (1967), 205.

14. R. Hermann, J. Math. Phys. 11 (1970), 182.

15. R. Hermann, Proceedings of the 1969 Battelle-Seattle Rencontres, Springer-Verlag (1970), p. 312.

16. R. Hermann, Lie algebras and quantum mechanics, W.A. Benjamin, N.Y., 1970.

17. F.J. Bloore, M. Assimakapoulos, and I.R. Ghobrial, Quantization as a deformation theory, preprint, Dept. of Applied Math., Univ. of Liverpool.

II.

QUANTUM AND FERMION DIFFERENTIAL GEOMETRY A LA CARTAN

Abstract

A new differential geometric formalism is described which can serve as the foundation of quantum-elementary particle physics. For example, it covers the recent fermion space-supersymmetry work and gives a precise differential-geometric meaning to the "quantum" Hamiltonian equations of motion.

1. Introduction

Recent work[1,2,3,4,5] on supersymmetry, fermion spaces, graded Lie algebras,... has suggested the need for an extension of standard differential geometry. In particular, F. Berezin and G. Kac[1] have extended the notion of "Lie group". Berezin and Marinov[3] have adapted the ideas of classical mechanics. B. Zumino (in reference 4, p. 265) has briefly suggested an extension of the usual notion of "differential form" to "fermion" (i.e., Grassman algebra-valued) differential forms. In this note I will sketch a way to do this so as to cover both this material and the differential-geometric foundations of quantum mechanics. This will be done by introducing "differential forms" associated with an arbitrary associative algebra $\mathscr{A}$. Many of the ideas of E. Cartan[6,7] (which are at the foundation of the modern work in differential geometry, and which often have profound physical ramifications) carry over to this formalism. For example, in this note I will show how to develop particle and field mechanics into this general framework.

I will assume that the reader is familiar with the elementary ideas of modern differential geometry,[7] particularly the part relating to differential forms, and with elementary[8] ideas of algebra, e.g., the notion of "algebra", "associativity", etc.

2. Differential Forms for an Associative Algebra

Let $\mathscr{A}$ be an associative algebra, say, with the real or complex numbers as field of scalars, and with a unit element "1". We define vector spaces, labelled by an integer $r = 0, 1, 2, \ldots$ and denoted by $\mathscr{D}^r(\mathscr{A})$, called the r-<u>th degree differential forms over</u> $\mathscr{A}$. We also define an "exterior product"

$$(\omega_1, \omega_2) \to \omega_1 \wedge \omega_2 \quad,$$

mapping $\mathscr{D}^r(\mathscr{A}) \times \mathscr{D}^s(\mathscr{A})$ into $\mathscr{D}^{r+s}(\mathscr{A})$, and an "exterior derivative" map

$$d: \mathscr{D}^r(\mathscr{A}) \to \mathscr{D}^{r+1}(\mathscr{A})$$

such that

$$dd = 0 \quad. \tag{2.1}$$

First, $\mathscr{D}^0(\mathscr{A}) = \mathscr{A}$. We shall begin with $r = 1$ and 2, $\mathscr{D}^1(\mathscr{A})$ consists of linear combinations of elements of the form

$$A_1 (dA_2) A_3$$

for $A_1, A_2, A_3 \in \mathscr{A}$,

with certain relations introduced.

$$d(A_1 A_2) = (dA_1) A_2 + A_1 (dA_2) \tag{2.2}$$

for $A_1, A_2 \in \mathscr{A} \quad .$

(Also, the more-or-less obvious relation induced by (2.2) and the associative law.) Two-forms are linear combinations of objects of the type

$$A_1(dA_2)A_3 \wedge (dA_4)A_5 \ .$$

The exterior derivative operation d is defined as $\mathcal{D}^1(\mathcal{A})$ as follows:

$$d(A_1 dA_2 A_3) = dA_1 \wedge dA_2 A_3 - A_1 dA_2 \wedge dA_3 \qquad (2.3)$$

((2.3) is required for consistency with (2.1).) Exterior product is defined as follows:

$$(A_1 dA_2 A_3) \wedge (A_4 dA_5 A_6) = A_1(dA_2)A_3 A_4 \wedge (dA_5)A_6 \qquad (2.4)$$

If $\mathcal{A}$ satisfies certain commutative relations, their exterior derivatives must also be introduced as relations. For example, suppose that A_1, A_2 satisfy:

$$A_1 A_2 \pm A_2 A_1 = c1 \ , \qquad (2.5)$$

where c belongs to the scalar field. Apply d to both sides, and use (2.2):

$$(dA_1)A_2 + A_1 dA_2 \pm ((dA_2)A_1 + A_2 dA_1) = 0 \qquad (2.6)$$

Now, one wants dA_1, dA_2 to be <u>independent</u>. This requires that

$$(dA_1)A_2 = \pm (dA_2)A_1 \qquad (2.7)$$

Apply d:

$$(dA_1) \wedge (dA_2) = \mp (dA_2) \wedge dA_1 \tag{2.8}$$

This is the basic commutation relation in this generalized differential form algebra.

3. **Boson Quantum Differential Forms**

Now, let $\mathcal{A}$ be generated by elements (P_i, Q^j), $1 \le i, j \le n$, satisfying the **Heisenberg commutation relations**, i.e.,

$$P_i Q^j - Q^j P_i = \sqrt{-1}\, \delta_i^j\, \hbar \tag{3.1}$$

$$P_i P_j - P_j P_i = 0 = Q^i Q^j - Q^j Q^i \tag{3.2}$$

Applying d, one obtains the following commutative relations:

$$A_1 dA_2 = dA_2 A_1 \tag{3.4}$$

$$dA_1 \wedge dA_2 = -dA_2 \wedge dA_1 , \tag{3.5}$$

for $A_1, A_2 \in \mathcal{A}$,

which are, of course, the same as for the abelian case. In other words, the "differentials" of "quantum observables" obey the same algebraic rules as those of classical observables. The elements of $\mathcal{D}^r(\mathcal{A})$ will be called (Boson) quantum differential forms.

4. Hamiltonian Particle Mechanics

Using Cartan's methods[6] Hamiltonian mechanics is essentially defined by means of a two-differential form. Having defined differential forms for general associative algebras, one can attempt to define Hamiltonian mechanics within that context. In fact, there are methods[10,11,12] available for treating field theories in Cartan's spirit using higher degree differential forms.

Suppose that $\mathscr{A}$ contains an element "t" (to be identified with physical "time", of course) such that

$$tA = At \tag{4.1}$$

for all $A \in \mathscr{A}$.

Then,

$$tdA = dAt \tag{4.2}$$

$$dt \wedge dA = -dA \wedge dt \tag{4.3}$$

Suppose that Ω is an element of $\mathscr{D}^2(\mathscr{A})$. A set (ω_i, ω^i), $1 \le i, j \le n$, of one-forms is said to be <u>canonical</u> with respect to Ω if

$$\Omega = \omega_i \wedge \omega^i . \tag{4.4}$$

The <u>characteristic curves of</u> Ω are those curves on which

$$\omega_i = \omega^i = 0 \tag{4.5}$$

When made explicit in terms of physical observables, they turn out to be the <u>dynamical equations of motion</u>.

To see the meaning of these equations, consider two cases.

a) $\mathscr{A}$ = the Grassman algebra generated by (θ^i) with the variable "t" added.

$$\Omega = s_{ij} d\theta^i \wedge d\theta^j - dh \wedge dt \tag{4.6}$$

$$d\theta^i \wedge d\theta^j = d\theta^j \wedge d\theta^i$$

where (s_{ij}) is a symmetric, constant non-degenerate matrix, and h is an element of $\mathscr{A}$. Let h_i, h_t be the elements of $\mathscr{A}$ such that:

$$dh = h_i d\theta^i + h_t dt \tag{4.7}$$

(Thus, $h_i = \partial h/\partial \theta^i$; $h_t = \partial h/\partial t$.) In order to achieve (6.4), write:

$$\omega_i = a_{ij}(d\theta^j - h^j dt)$$

$$\omega^i = a^i_j(d\theta^j - h^j dt) ,$$

with (a_{ij}), (a^i_j) constant, h^j elements of $\mathscr{A}$ to be determined.

$$\omega_i \wedge \omega^i = a_{ij}(d\theta^j - h^j dt) \wedge a^i_k(d\theta^k - h^k dt)$$

$$= \frac{1}{2}(a_{ij} a^i_k + a_{ik} a^i_j) d\theta^j \wedge d\theta^k$$

$$+ a_{ij} a^i_k (h^j d\theta^k \wedge dt - d\theta^j h^k \wedge dt)$$

Comparing this with (6.6) and (6.7), we have:

$$s_{ij} = \frac{1}{2}(a_{ij} a_k^i + a_{jk} a_j^i) \qquad (4.8)$$

$$h_k d\theta^k = a_{ij} a_k^i (h^j d\theta^k - d\theta^j h^k) \qquad (4.9)$$

In order to determine the rest, we make an assumption about the "Hamiltonian" h. The simplest assumption seems to be the following one :

$$h \in \mathscr{A}^+ \quad ;$$

$$h_i, h^i \in \mathscr{A}^- \quad .$$

Thus,

$$h_k = a_{ij} a_k^i h^j + a_{ik} a_j^i h^j \qquad (4.10)$$

For example, the simplest case would be:

$$s_{ij} = \delta_{ij} \quad .$$

Then,

$$a_{ij} = \delta_{ij} = \delta_j^i$$

$$h_k = 2h^j \quad .$$

The dynamic equations (6.5) are then:

$$\frac{d\theta^j}{dt} = h^j(\theta(t)) \tag{4.11}$$

These are precisely the "classical spinning particle" equations. Note that our formalism for this case essentially coincides with that of Berezin and Marinov.[13]

b) $\mathcal{A}$ = Quantum mechanical algebra, described in Section 5, generated by P_i, Q^i satisfying the Heisenberg commutation relations.

Suppose $\mathcal{A}$ also contains "t", which commutes with everything. Set:

$$\Omega = dP_i \wedge dQ^i - dh \wedge dt \tag{4.12}$$

$$dh = h_i dQ^i + h^i dP_i + h_t dt \tag{4.13}$$

(Thus, $h_i = \partial h/\partial Q^i$; $h^j = \partial h/\partial P^j$; $h_t = \partial h/\partial t$.) Then,

$$\Omega = (dP_i + h_i dt) \wedge (dQ^i - h^i dt) \tag{4.14}$$

The equations of motion (6.5) are then

$$\frac{dP_i}{dt} = \frac{\partial h}{\partial Q^i} \quad ; \quad \frac{dQ^i}{dt} = -\frac{\partial h}{\partial P^i} \tag{4.15}$$

These are the <u>quantum mechanical version of Hamilton's equations of classical mechanics</u>. Notice that they are defined completely <u>independently of classical mechanics</u>, in a rather "intrinsic" way.

There is no need to mention the problem of "quantizing" a classical Hamiltonian; once the Hamiltonian h is given as a quantum observable, everything is determined by the differential-geometric formalism

5. Lagrangian Field Theories

As another illustration of the physical applications of the formalism, I will now extend the work of Berezin and Marinov[13] concerning Lagrangian particle mechanics to cover Lagrangian field theories using the differential-geometric methods [10] developed for "classical" Lagrangian field theories.

Let the algebra $\mathcal{A}$ be generated by elements $(x^\mu, \varphi^a, \varphi^a_\mu)$, $0 \leq \mu \leq 3$, $1 \leq a \leq m$, such that:

$$dx^\mu \wedge dx^\nu = -dx^\nu \wedge dx^\mu$$

$$dx^\mu \wedge d\varphi^a = -d\varphi^a \wedge dx^\mu$$

$$d\varphi^a \wedge d\varphi^b = d\varphi^b \wedge d\varphi^a$$

etc.

(Thus, the "space" associated with the variables (x, φ) --which actually does not exist in the usual mathematical sense--is what the physicists call "superspace".[2,4] The space of variables $(x, \varphi^a, \varphi^a_\mu)$ is its "jet space".) Let $L(x, \varphi)$ be an element of $\mathcal{A}$, with $dx = dx^0 \wedge \cdots \wedge dx^4$. Associate to the "variational problem"

$$\int L \, dx$$

the "Cartan form"

$$\theta = L\,dx + L_a^\mu \left(\frac{\partial}{\partial x^\mu} \,\lrcorner\, dx\right) \wedge (d\varphi^a - \varphi_\mu^a\,dx^\nu) \quad,$$

where

$$dL = L_\mu\,dx^\mu + L_a\,d\varphi^a + L_a^\mu\,d\varphi_\mu^a \quad.$$

Then, one readily verifies that

$$d\theta = \omega_a \wedge (d\varphi^a - \varphi_\nu^a\,dx^\nu)$$

where

$$\omega_a = L_a\,dx - d(L_a^\mu) \wedge \left(\frac{\partial}{\partial x^\mu} \,\lrcorner\, dx\right)$$

The dynamical equations are then

$$\omega_a = 0 = (d\varphi^a - \varphi_\nu^a\,dx^\nu) \quad.$$

This formalism can be adapted to cover all sorts of constraints, higher-derivative Lagrangians, etc.

Bibliography

1. F. Berezin and G.I. Kac, Mat. Sb. USSR, 82 (1970), 124 Eng. Trans. 1970 (11), 311.

2. J. Wess and B. Zumino, Nucl. Phys. B70, 39 (1974).

3. L. Corwin, Y. Ne'eman and S. Sternberg, Reviews of Modern Physics 47 (1975), 573.

4. R. Arnowitt and P. Nath (eds.), Gauge Theories and Modern Field Theory, MIT Press, 1976.

5. B. Kostant, Graded Manifolds, Graded Lie Theory and Pre-Quantization, preprint, MIT Mathematics Department.

6. E. Cartan, Leçons sur les Invariants Intégraux, Hermann, Paris, 1971.

7. E. Cartan, Les Systèms Differentials Extérieurs et leurs Applications Géometrique, Hermann, Paris, 1971.

8. R. Hermann, Differential Geometry and the Calculus of Variations, Academic Press, 1968.

9. R. Hermann, General Algebraic Ideas (Interdisciplinary Mathematics, Vol. 1), Math Sci Press, Brookline, Mass., 1973.

10. R. Hermann, Lie Algebras and Quantum Mechanics, W.A. Benjamin, 1970.

11. R. Hermann, Vector Bundles in Mathematical Physics, Parts I and II, W.A. Benjamin, 1970.

12. R. Hermann, Geometry, Physics and Systems, Dekker, New York, 1973.

13. F. Berezin and M. Marinov, Particle Spin Dynamics as the Grassman Variant of Classical Mechanics, preprint of the Institute of Theoretical and Experimental Physics, Moscow, 1976.

III.

GENERALIZED BEREZIN-KAC GROUPS IN QUANTUM DIFFERENTIAL GEOMETRY

Abstract

Ideas of Berezin and Kac on the definition of generalized Lie groups are further algebracized and extended to "quantum" situations.

1. Introduction

There has been much interest lately among physicists and mathematicians in creating a general differential geometric formalism which will unify the theory of bosons and fermions. In [1] I suggested a way of doing this (with a bonus of a possible new "quantum" differential geometry) starting with an arbitrary associative algebra. (The commutative case leads to "bosons". The Grassman-variant of commutativity leads to "fermions". Commutation relations such as those of Heisenberg or Clifford lead to "quantum" versions of these differential forms.)

Berezin and Kac have developed [2] a theory of generalized "Lie groups" and "homogeneous spaces" based on what physicists now call "superspaces". In terms of [1] this involves working with an algebra which satisfies the Grassman variant of "commutativity". In this note, I will sketch how to extend their ideas to more general associative algebras. In particular, if applied to the associative algebras defined by quantum mechanical observables, these objects might be useful in physics.

2. Generalizations of Lie Groups Based on Associative Algebras

Let $\mathcal{A}$ be an associative algebra, with the real or complex numbers as field of scalars. Suppose $\mathcal{A}$ has a unit element, labelled as "1". Let (x^i), $1 \leq i, j \leq n$, be a set of elements of $\mathcal{A}$ which are linearly independent over the scalars, do not contain the scalar element, but generate $\mathcal{A}$, i.e., every element of $\mathcal{A}$ can be written as the sum of scalars and a polynomial in the x's. For the moment, we leave open the commutation relations satisfied by these elements.

Let $\mathcal{A} \otimes \mathcal{A}$ be the tensor product [3] of the <u>vector spaces</u> $\mathcal{A}$ and $\mathcal{A}$. We will suppose an associative algebra structure is chosen for $\mathcal{A} \otimes \mathcal{A}$, i.e., a linear map

$$\mathcal{A} \otimes \mathcal{A} \otimes \mathcal{A} \otimes \mathcal{A} \to \mathcal{A} \otimes \mathcal{A} \quad ,$$

such that $1 \otimes 1$ is a unit. (Two choices, one "boson", the other "fermion" are possible.)

Let γ be a linear map

$$\mathcal{A} \to \mathcal{A} \otimes \mathcal{A}$$

Let $\gamma \otimes 1$, $1 \otimes \gamma$ be linear maps: $\mathcal{A} \otimes \mathcal{A} \to \mathcal{A} \otimes \mathcal{A} \otimes \mathcal{A}$ such that:

$$(\gamma \otimes 1)(a_1 \otimes a_2) = \gamma(a_1) \otimes a_2$$

$$(1 \otimes \gamma)(a_1 \otimes a_2) = a_1 \otimes \gamma(a_2)$$

for $a_1, a_2 \in \mathcal{A}$.

<u>Definition.</u> γ is said to be <u>multiplicative</u> if it is a homomorphism of the algebraic structures on $\mathcal{A}$ and $\mathcal{A} \otimes \mathcal{A}$. γ is said to be <u>co-associative</u> [4] if it satisfies the following condition:

$$(\gamma \otimes 1)\gamma = (1 \otimes \gamma)\gamma \qquad (2.1)$$

Definition. γ is said to define a <u>generalized Lie group structure</u> based on $\mathscr{A}$ if it is multiplicative and co-associative.

We will not work with the most general possible such γ, but will restrict attention to those that have the following form:

$$\gamma(a) = A_0(a) \otimes 1 + A_i(a) \otimes x^i + A_{ij}(a) \otimes x^i x^j + \cdots \qquad (2.2)$$

for all $a \in \mathscr{A}$,

where $A_0, A_i, A_{ij}, \ldots$ are linear maps $\mathscr{A} \to \mathscr{A}$. This is a general form of formulas written down by Berezin and Kac [2]. The conditions that the $A_0, A_i, A_{ij}, \ldots$ must satisfy in order that γ be multiplicative and co-associative turn out to be significant, although very complicated. In the case of a traditional Lie group situation (i.e., where the x^i satisfy boson commutative relations) the A_i turn out to be generators of the Lie algebra of the group and the $A_{ij}, A_{ijk}, \ldots$ turn out to be elements in its universal enveloping algebra. In the Berezin-Kac case [2], the x^i satisfy some commutation, some anti-commutation relations, and the A_i turn out to generate a graded Lie algebra.

As a variant of these ideas, we now consider a generalization which might be significant for quantum mechanics.

3. <u>Boson Quantum Lie Groups</u>

Suppose that generators x^i of $\mathscr{A}$ satisfy only the following relations:

$$x^i x^j - x^j x^i = \hbar \lambda^{ij} 1 \quad , \tag{3.1}$$

(λ^{ij}) scalars, $\hbar$ = Planck's constant .

$\mathcal{A}$ might be called the <u>Heisenberg associative algebra</u>. Let us adopt the following "boson" rule for multiplication in $\mathcal{A} \times \mathcal{A}$.

$$(a_1 \otimes a_2)(b_1 \otimes b_2) = (a_1 b_1 \otimes a_2 b_2) \tag{3.2}$$

We shall write down the low-order conditions that γ, given by (2.2), be multiplicative:

$$\gamma(a_1)\gamma(a_2) = (A_0(a_1) \otimes 1 + A_i(a_1) \otimes x^i + A_{ij}(a_1) x^i x^j + \ldots)(A_0(a_2) \otimes 1 + \ldots$$

$$= A_0(a_1)A_0(a_2) \otimes 1 + (A_0(a_1)A_i(a_2) + A_i(a_1)A_0(a_2)) \otimes x^i$$

$$+ (A_i(a_1)A_j(a_2) + A_0(a_1)A_{ij}(a_2) + A_{ij}(a_1)A_0(a_2)) \otimes x^i x^j$$

$$+ \ldots$$

$$= A_0(a_1 a_2) \otimes 1 + A_i(a_1 a_2) \otimes x^i + A_{ij}(a_1 a_2) \otimes x^i x^j + \ldots$$

$$\tag{3.3}$$

We can suppose without loss in generality that:

$$A_{ij} = A_{ji} \quad . \tag{3.4}$$

Now,

$$A_i(a_1)A_j(a_2) \otimes x^i x^j = \tfrac{1}{2}(A_i(a_1)A_j(a_2) \otimes x^i x^j + A_j(a_1)A_i(a_2) \otimes x^j x^i)$$

$$= \text{, using (3.1),}$$

$$\tfrac{1}{2}(A_i(a_1)A_j(a_2) \otimes x^i x^j + A_j(a_1)A_i(a_2) \otimes (x^{ij} + \hbar \lambda^{ij} 1)$$

$$= \frac{1}{2}(A_i(a_1)A_j(a_2) + A_j(a_1)A_i(a_2)) \otimes x^i x^j + \hbar \lambda^{ij} A_j(a_1)A_i(a_2) \otimes 1$$

(3.5)

Let us now (as a sample calculation) insert (3.5) into (3.3), cut off the terms indicated by ..., and equate coefficients of both sides. The result is the following relation:

$$A_{ij}(a_1 a_2) = A_{ij}(a_1)A_0(a_2) + A_0(a_1)A_{ij}(a_2) + \frac{1}{2}(A_i(a_1)A_j(a_2) + A_j(a_1)A_i(a_2))$$

(3.6)

$$A_i(a_1 a_2) = A_i(a_1)A_0(a_2) + A_0(a_1)A_i(a_2) \tag{3.7}$$

$$A_0(a_1 a_2) = A_0(a_1)A_0(a_2) - \hbar \lambda^{ij} A_i(a_1)A_j(a_2) \quad . \tag{3.8}$$

In the "classical limit", $\hbar \to 0$, these equations become uncoupled, and can be solved in the following form:

$$A_0^{(0)}(a) = a \quad .$$

$A_i^{(0)}$ is a derivation of the algebra $\mathscr{A}$, i.e., a first order, linear differential operator in the variables x^i. $A_{ij}^{(0)}$ is a second order linear differential operator in the variables x^j, and so forth. The A_i then generate (under commutation) a Lie algebra in the usual way. It is natural to try a perturbation solution:

$$A_0(a) = a + \hbar A_0^{(1)}(a) + \cdots$$

$$A_i(a) = A_i^{(0)}(a) + \hbar A_i^{(1)}(a) + \cdots$$

The details of the explicit solutions of these equations will be presented in another publication. The equations for co-associativity are even more complicated.

4. Generalized Homogeneous Spaces

Berezin and Kac [2] have already indicated how Lie group homogeneous spaces may be generalized; their ideas may be readily carried over. Let $\mathcal{A}, \mathcal{B}$ be associative algebras with associative algebraic structures also prescribed for $\mathcal{A} \otimes \mathcal{A}$ and $\mathcal{A} \otimes \mathcal{B}$. Let

$$\gamma: \mathcal{A} \to \mathcal{A} \otimes \mathcal{A}$$

Define $\mathcal{A}$ as a generalized Lie group, i.e., γ is multiplicative and co-associative. Let

$$\delta: \mathcal{B} \to \mathcal{A} \otimes \mathcal{B} \tag{4.1}$$

be a linear map. δ defines $\mathcal{B}$ as a <u>generalized homogeneous space</u> of $(\mathcal{A}, \gamma)$ if δ is multiplicative, i.e., is an algebra homomorphism, and if it satisfies the following generalization of the co-associative law:

$$(1 \otimes \delta)\delta = (\gamma \otimes 1)\delta \quad . \tag{4.2}$$

Bibliography

1. R. Hermann, "Quantum and Fermion Differential Geometry à la Cartan", preprint, Harvard University Physics Department, 1977.

2. F. Berezin and G.I. Kac, MAT. SB. USSR, 82, 124 (1970); English translation, 1970 (11), 311.

3. W. Greub, Multilinear Algebra, Springer-Verlag, Berlin, 1967.

4. M. Sweedler, Hopf Algebras, W.A. Benjamin, 1969.

IV.

THE DIFFERENTIAL-GEOMETRIC FOUNDATIONS OF "SUPERSPACE" AND "SUPERSYMMETRY" THEORIES, I

GRADED LIE ALGEBRAS AND THEIR GEOMETRIC REPRESENTATIONS

Abstract

Graded Lie algebras are first discussed as tensor products of Grassman algebras and Lie algebras (i.e., as generalized "current algebras"), then as "vector fields" on "Fermion spaces", then as associated with combined symplectic and Riemannian structures on the usual sort of manifolds.

1. INTRODUCTION

This series of papers attempts to bring together two threads running through the mathematical physics literature in recent years. On the one hand, "current algebras" have been defined [1-9] as Lie algebras, constructed from commutative associative algebras and Lie algebras by tensor products and certain "twisting" operations that physicists call "Schwinger terms". On the other hand, there has been much more extensive work both by physicists and mathematicians on "graded Lie algebras", in the physical context of "supersymmetries" [10, 11,12]. In fact, these physical ideas have suggested the construction of a "differential geometry" based on "fermion spaces" which generalizes traditional differential geometry. This is done by exploiting the fact that differential geometry on a manifold can be completely described in algebraic terms by means of algebraic properties of the (commutative associative) algebras of C^∞, real-valued functions on the manifold. One then replaces this function algebra by a Grassman algebra, and develops differential geometric ideas by analogy.

In this paper, I will approach "graded Lie algebras" by analogy with the way "current algebras" were treated in my previous work [1-9]. Although this paper is mathematical, it is directed towards exploring certain territory that might be useful for the applications to elementary particle physics.

Another topic treated here is an alternate geometric framework for realizing Graded Lie Algebras--in terms of a pair consisting of a symmetric and skew-symmetric differential form on a manifold. This procedure generalizes the natural way Graded Lie Algebras arise in quantum mechanics--as Hermitian operators with the bracket given by commutator or anticommutator. Relations with quantum field theory are also briefly discussed.

In this paper it is assumed that the reader is familiar with certain standard mathematical facts and ideas concerning differential geometry on manifolds, linear and multilinear algebra, etc.

2. LIE ALGEBRAS OF THE SIMPLEST "CURRENT ALGEBRA" TYPE, WITH NO SCHWINGER TERMS

All algebras, vector spaces, etc. will be taken with the real numbers as the field of scalars, unless mentioned otherwise. Similarly, "tensor products" of vector spaces will be defined relative to this field of scalars, unless mentioned otherwise. The standard mathematical facts about linear and multilinear algebra on vector spaces is assumed to be known [13,14].

Let us first recall how the simplest sort of "current algebra" will be defined algebraically. Let $\underset{\sim}{L}$ be a Lie algebra and let $\mathscr{F}$ be a commutative associative algebra. Consider the tensor product

$$\mathscr{F} \otimes \underset{\sim}{L} \equiv \underset{\sim}{L}_{\mathscr{F}} \quad .$$

Define a bilinear operation on it as follows:

$$[f_1 \otimes A, f_2 \otimes B] = f_1 f_2 \otimes [A,B] \tag{2.1}$$

for $f_1, f_2 \in \mathcal{F}$; $A, B \in \underset{\sim}{L}$.

It is readily verified that this is a Lie algebra structure on $\underset{\sim}{L}_{\mathcal{F}}$.

This is the simplest sort of mathematical object of the type that physicists call a "current algebra". Here is how it may be identified with more familiar objects to the phsyicist. Suppose $\mathcal{F}$ are the C^∞, real-valued functions on R^3. Denote a point of R^3 by $\vec{x}$. Suppose for the moment that the "Dirac delta function", $\delta_{\vec{x}}$, belonged to $\mathcal{F}$. Set:

$$A(\vec{x}) = \delta_{\vec{x}} \otimes A \tag{2.2}$$

for $\vec{x} \in R^3$, $A \in \underset{\sim}{L}$.

Then, the defining relation (2.2) is equivalent (using the customary rules for distributions-generalized functions) to a relation the physicist would write in the following form:

$$[A(\vec{x}), B(\vec{y})] = [A,B](\vec{x})\delta(\vec{x}-\vec{y}) . \tag{2.3}$$

This would be even more familiar to a physicist if a basis (A^a), $1 \leq a,b \leq m$, were chosen for $\underset{\sim}{L}$, and (2.3) were written as

$$[A^a(\vec{x}), A^b(\vec{y})] = C_c^{ab} A^c(\vec{x})\delta(\vec{x}-\vec{y}) , \tag{2.4}$$

where (C_c^{ab}) are the structure constants of the Lie algebra $\underset{\sim}{L}$.

In fact, this is often identified with a "quantum field theoretic" relation of the form:

$$[V_0^a(\vec{x}), V_0^b(\vec{y})] = iC_c^{ab}V_0^c(\vec{x})\delta(\vec{x}-\vec{y}) \qquad (2.5)$$

where $(V_u^a(x))$ are "currents". Typically, a will be an "internal symmetry" index. Thus, we see that there are simple ways to express physical relations like (2.4) or (2.5) in precise algebraic terms. Relation (2.5) is the simplest of a sort of relation that one sees often in the relevant physical literature (often with "Schwinger terms") and much of it can be handled in a similar algebraic way by means of techniques developed in [1-9]. This material is mentioned here as motivation--we next want to replace $\mathscr{F}$ by a Grassman algebra, obtaining a mathematical object which is more general than a Lie algebra.

3. THE TENSOR PRODUCT OF A GRASSMAN ALGEBRA AND A LIE ALGEBRA. THE DEFINITION OF A GRADED LIE ALGEBRA

First, we shall define what we mean by a "Grassman algebra" $\mathscr{G}$, since it is not such a commonly encountered object as a "Lie algebra". To define it, one should be given, first, a vector space structure for $\mathscr{G}$, and, second, a bilinear multiplication operation

$$\mathscr{G} \times \mathscr{G} \to \mathscr{G} ,$$

denoted by

$$(g_1, g_2) \to g_1 \wedge g_2 \quad ,$$

and a direct sum decomposition

$$\mathcal{G} = \mathcal{G}^0 \oplus \mathcal{G}^1 \oplus \cdots$$

of $\mathcal{G}$ into subspaces indexed by a nonnegative integer, such that:

$$\mathcal{G}^n \wedge \mathcal{G}^m \subset \mathcal{G}^{n+m} \tag{3.1}$$

$$g_1 \wedge g_2 = (-1)^{nm} g_2 \wedge g_1 \tag{3.2}$$

$$\text{for } g_1 \in \mathcal{G}^n, \quad g_2 \in \mathcal{G}^m \quad .$$

Here is a useful convention to simplify writing relations like (3.2). Write it as

$$g_1 \wedge g_2 = (-1)^{|g_1||g_2|} g_2 \wedge g_1 \tag{3.3}$$

where $g_1 \to |g_1| = n$ is the "gradation" which g_1 carries along. When we write a relation like (3.3) we shall leave off the statement "for $g_1 \in \mathcal{G}^n$, $g_2 \in \mathcal{G}^m$", implicitly understanding that the relation should hold only for the elements of $\mathcal{G}$ which belong to the "grading" subspaces $\mathcal{G}^0, \mathcal{G}^1, \ldots$ (The "exterior algebra" [13,14] of a vector space, of course, provides the prototype, hence it is reasonable to choose general notation to agree as far as possible with the traditional notation used for exterior algebras.)

Now, let $\mathcal{G}$ be such a Grassman algebra, and let $\underset{\sim}{L}$ be another vector space with an algebraic operation

$$\underset{\sim}{L} \times \underset{\sim}{L} \to \underset{\sim}{L} \quad,$$

that we denote by [,] and about which we, for the moment, make no other assumptions. As before, set:

$$\underset{\sim}{L}_{\mathcal{G}} = \mathcal{G} \otimes \underset{\sim}{L} \tag{3.4}$$

Define a bilinear map

$$\underset{\sim}{L}_{\mathcal{G}} \times \underset{\sim}{L}_{\mathcal{G}} \to \underset{\sim}{L}_{\mathcal{G}} \quad,$$

which we also denote by [,], by means of the following formula:

$$[g_1 \otimes A, \, g_2 \otimes B] = (g_1 \wedge g_2) \otimes [A,B] \tag{3.5}$$

Let us now see under what conditions this operation on $\underset{\sim}{L}_{\mathcal{G}}$ is "commutative" or "graded commutative".

$$[g_2 \otimes B, \, g_1 \otimes A] = (g_2 \wedge g_1) \otimes [B,A]$$

$$= (-1)^{|g_1||g_2|} (g_1 \wedge g_2) \otimes [B,A]$$

$$= \pm (-1)^{|g_1||g_2|} [g_1 \otimes A, \, g_2 \otimes B] \tag{3.6}$$

if

$$[A,B] = \pm [B,A] \quad . \tag{3.7}$$

Thus, if the bracket operation on $\underset{\sim}{L}$ is "commutative" (e.g., if it is a Jordan algebra), i.e., (3.7) holds with a plus sign, then $\underset{\sim}{L}_{\mathcal{G}}$ is "graded commutative". Similarly, if $\underset{\sim}{L}$ is "skew-commutative" (e.g., a Lie algebra) then $\underset{\sim}{L}_{\mathcal{G}}$ is "graded skew commutative".

Now, let us suppose that $\underset{\sim}{L}$ is a Lie algebra, i.e., the bracket on it is skew-commutative and satisfies the Jacobi identity:

$$[A,[B,C]] = [[A,B],C] + [B,[A,C]] \qquad (3.8)$$

for $A, B, C \in \underset{\sim}{L}$.

Let us see how the Jacobi identity for $\underset{\sim}{L}$ is reflected in terms of $\underset{\sim}{L}_{\mathcal{G}}$.

$$[g_1 \otimes A, [g_2 \otimes B, g_3 \otimes C]] = [g_1 \otimes A, (g_2 \wedge g_3) \otimes [B,C]]$$

$$= (g_1 \wedge g_2 \wedge g_3) \otimes [A, [B,C]]$$

Similarly,

$$[[g_1 \otimes A, g_2 \otimes B], g_3 \otimes C] = g_1 \wedge g_2 \wedge g_3 \otimes [[A,B],C]$$

$$[g_2 \otimes B, [g_1 \otimes A, g_3 \otimes C]] = [g_2 \otimes B, (g_1 \wedge g_3) \otimes [A,C]]$$

$$= g_2 \wedge g_1 \wedge g_3 \otimes [B, [A,C]]$$

$$= (-1)^{|g_1||g_2|} g_1 \wedge g_2 \wedge g_3 \otimes [B, [A,C]]$$

Thus, we have the following basic identity:

$$[g_1 \otimes A, [g_2 \otimes B, g_3 \otimes C]] = [[g_1 \otimes A, g_2 \otimes B], g_3 \otimes C]$$
$$+ (-1)^{|g_1||g_2|} [g_2 \otimes B, [g_1 \otimes A, g_3 \otimes C]] \quad (3.9)$$

This is called the <u>Graded Jacobi Identity</u>.

Let us reinterpret (3.6) and (3.9) in the following way. Set:

$$\underset{\sim}{L}{}^n_{\mathcal{G}} = \mathcal{G}^n \otimes \underset{\sim}{L} \quad (3.10)$$

for $n = 0, 1, \ldots$

This defines a grading of $\underset{\sim}{L}_{\mathcal{G}}$, i.e., a direct sum decomposition into linear subspaces labelled by a nonnegative integer. Define a bracket operation on $\underset{\sim}{L}_{\mathcal{G}}$ by means of formula (3.5). It satisfies the following rules:

$$[\underset{\sim}{L}{}^n_{\mathcal{G}}, \underset{\sim}{L}{}^m_{\mathcal{G}}] \subset \underset{\sim}{L}{}^{n+m} \quad (3.11)$$

$$[\alpha, \beta] = (-1)^{|\alpha||\beta|} [\beta, \alpha] \quad , \quad (3.12)$$

for $\alpha \in \underset{\sim}{L}{}^n$, $\beta \in \underset{\sim}{L}{}^m$, with $n = |\alpha|$, $m = |\beta|$

$$[\alpha, [\beta, \gamma]] = [[\alpha, \beta], \gamma] + (-1)^{|\alpha||\beta|} [\beta, [\alpha, \gamma]] \quad . \quad (3.13)$$

This leads into a general

<u>Definition</u>. A <u>graded Lie algebra</u> structure on a vector space GL is defined by the following algebraic structure:

a) A bilinear map $GL \times GL \to GL$, denoted by $[\ ,\ ]$ and called the __bracket__.

b) A grading of GL, i.e., a direct sum decomposition into linear subspaces GL^n labelled by a non-negative integer.

The following rules must be satisfied:

$$[GL^n, GL^m] \subset GL^{n+m} \qquad (3.14)$$

$$[\alpha, \beta] = (-1)^{|\alpha||\beta|} [\beta, \alpha] \qquad (3.15)$$

for $\alpha, \beta \in GL$.

$$[\alpha, [\beta, \gamma]] = [[\alpha, \beta], \gamma] - (-1)^{|\alpha||\beta|} [\beta, [\alpha, \gamma]] \qquad (3.16)$$

Thus, we have proved:

__Theorem 3.1__. The tensor product of a Grassman algebra and a Lie algebra is a graded Lie algebra.

4. LINEAR REPRESENTATIONS OF A GRADED LIE ALGEBRA

Let GL be a graded Lie algebra, and let V be a vector space. Let $L(V,V)$ denote the vector space of linear maps $V \to V$.

__Definition__. A linear representation of GL in V is a linear map

$$\rho: GL \to L(V,V)$$

such that:

$$\rho([\alpha,\beta]) = \rho(\alpha)\rho(\beta) - (-1)^{|\alpha||\beta|} \rho(\beta)\rho(\alpha) \qquad (4.1)$$

for $\alpha, \beta \in GL$.

This definition captures the prime <u>physical</u> reason for introducing Graded Lie Algebras. For $|\alpha|, |\beta|$ even, (4.1) is the <u>commutator</u>, for $|\alpha|, |\beta|$ odd, it is the <u>anticommutator</u>. Thus, it gives us the possibility of unifying "boson" and "fermion" commutation relations in a single algebraic structure.

Notice now that there is always one linear representation of a graded Lie algebra, namely the <u>adjoint representation</u>: Here,

$$V = GL \text{ itself}$$

$$\rho(\alpha)(v) = [\alpha, v] \qquad (4.2)$$

for $\alpha \in GL$, $v \in V$.

The Graded Jacobi Identity then guarantees that this is a linear representation.

We now turn to the study of another algebraic situation which leads to Graded Lie Algebras and their representations in a very natural way.

5. DERIVATIONS OF THE EXTERIOR ALGEBRA OF A VECTOR SPACE

Let V be a vector space. Suppose it is finite dimensional. Let V^d denote its dual. Let $\mathcal{G}(V)$ denote the <u>exterior algebra</u> defined on the vector space V^d, i.e., an element of $\mathcal{G}(V)$ is a skew-symmetric tensor in V^d. $\mathcal{G}(V)$ is graded by the degree of the tensor. Thus, an element $\omega \in \mathcal{G}(V)^n$ can be identified with an n-multilinear map

$$(v_1,\ldots,v_n) \to \omega(v_1,\ldots,v_n)$$

which is skew-symmetric. Exterior multiplication makes $\mathcal{G}(V)$ into a Grassman algebra in the usual way. (See [14,15] for a direct method of defining this algebra structure.)

Given $v \in V$, $\omega \in \mathcal{G}(V)^n$, let

$$v \lrcorner \omega \in \mathcal{G}(V)^{n-1}$$

be the <u>contraction</u> of ω by v, i.e.,

$$(v \lrcorner \omega)(v_2,\ldots,v_n) = \omega(v,v_2,\ldots,v_n)$$

for $v_2,\ldots,v_n \in V$.

Then, we have:

$$v \lrcorner (\omega_1 \wedge \omega_2) = (v \lrcorner \omega_1) \wedge \omega_2 + (-1)^{|\omega_1|} \omega_1 \wedge (v \lrcorner \omega_2) \tag{5.1}$$

Given $\alpha \in \mathcal{G}(V)^n$, $v \in V$, set:

$$\rho((\alpha,v))(\omega) = \alpha \wedge (v \lrcorner \omega) \quad . \tag{5.2}$$

Then

$$\rho((\alpha,v)(\omega_1 \wedge \omega_2)) = \alpha \wedge (v \lrcorner (\omega_1 \wedge \omega_2))$$

$$= \alpha \wedge (v \lrcorner \omega_1) \wedge \omega_2 + (-1)^{|\omega_1|} \omega_1 \wedge (v \lrcorner \omega_2))$$

$$= \rho((\alpha,v))(\omega_1) \wedge \omega_2$$

$$+ (-1)^{|\omega_1|}(-1)^{|\alpha||\omega_1|} \omega_1 \wedge \rho((\alpha,v)(\omega_2))$$

$$= \rho((\alpha,v)(\omega_1) \wedge \omega_2$$

$$+ (-1)^{(|\alpha|+1)|\omega_1|} \omega_1 \wedge \rho((\alpha,v)(\omega_2))$$

(5.3)

Suppose now that

$$\alpha \in \mathscr{G}(V)^n, \qquad \beta \in \mathscr{G}(V)^m,$$

$$v, v' \in V .$$

Let us compute the commutator and anti-commutator of $\rho((\alpha,v))$ and $\rho((\beta,v'))$.

$$\rho((\alpha,v))\rho((\beta,v'))(\omega) = \rho((\alpha,v)(\beta \wedge (v' \lrcorner \omega))$$

$$= \alpha \wedge (v \lrcorner (\beta \wedge (v' \lrcorner \omega)))$$

$$= \alpha \wedge [(v \lrcorner \beta) \wedge (v' \lrcorner \omega)$$

$$+ (-1)^{|\beta|} \beta \wedge (v \lrcorner v' \lrcorner \omega)]$$

Hence,

$$(\rho((\alpha,v)))\rho((\beta,v')) \pm \rho((\beta,v'))\rho((\alpha,v))(\omega)$$

$$= \alpha \wedge (v \lrcorner \beta) \wedge (v' \lrcorner \omega) + (-1)^{|\beta|} \alpha \wedge \beta \wedge (v \lrcorner v' \lrcorner \omega)$$

$$\pm \beta(v' \lrcorner \alpha) \wedge (v \lrcorner \omega) \mp (-1)^{|\alpha|} \beta \wedge \alpha \wedge (v' \lrcorner v \lrcorner \omega) \quad (5.4)$$

Let us choose the sign $\pm$ so that the second and fourth term on the right hand side of (5.4) add up to zero. This requires that:

$$(-1)^{|\beta|} = \pm (-1)^{|\alpha| + |\alpha||\beta|} ,$$

or

$$\pm = (-1)^{|\beta| + |\alpha| + |\alpha||\beta|} \quad (5.5)$$

Set:

$$|(\alpha,v)| = |\alpha| + 1 \quad (5.6)$$

$$|(\beta,v')| = |\beta| + 1 \quad (5.7)$$

$$|(\alpha,v)||(\beta,v')| = |\alpha||\beta| + |\alpha| + |\beta| + 1$$

Hence, (5.5) takes the form:

$$\pm = -(-1)^{|(\alpha,v)||(\beta,v')|}$$

We then have:

$$\rho((\alpha,v))\rho((\beta,v')) - (-1)^{|(\alpha,v)||(\beta,v')|} \rho((\beta,v'))\rho((\alpha,v))$$

$$= \rho((\alpha \wedge (v \lrcorner \beta),v')) - (-1)^{|(\alpha,v)||(\beta,v')|} \rho((\beta \wedge (v' \lrcorner \alpha),v)) \quad (5.8)$$

This suggests construction of a Graded Lie Algebra Structure on the vector space

$$\mathscr{G}(V) \otimes V$$

in such a way that

$$\alpha \otimes v \to \rho((\alpha,v))$$

is a linear representation. Namely, set:

$$|\alpha \otimes v| = |\alpha| + 1 \quad (5.9)$$

$$[\alpha \otimes v, \beta \otimes v'] = \alpha \wedge (v \lrcorner \beta) \otimes v' - (-1)^{|\alpha \otimes v||\beta \otimes v'|} \beta \wedge (v' \lrcorner \alpha) \otimes v \quad (5.10)$$

Let us sum up as follows

Theorem 5.1. These formulas define a Graded Lie Algebra structure on the vector space $\mathscr{G}(V) \otimes V$, together with a linear representation of it by linear maps on the underlying vector space of $\mathscr{G}(V)$.

Remark. The operations $\omega \to v \lrcorner \omega$, $\omega \to \alpha \wedge \omega$ are basically the <u>annihilation</u> and <u>creation operators</u> of quantum mechanics. This graded Lie algebra structure is also closely related to Clifford algebras constructed from V. This construction is also closely related to differential geometry based on "fermion" space. (See [10,11,12,16].) We now treat this point.

6. GRADED LIE ALGEBRAS AS "VECTOR FIELDS" IN FERMION SPACES

In modern differential geometry, all classical tensorial and spinorial analysis can be formulated purely "algebraically", replacing the underlying "space" or "manifold" by its commutative associative algebra of C^∞ functions. (See [14,18,19].) This theory can be developed in close analogy with the material of Section 5, as we shall now indicate.

Let V continue as a real finite dimensional vector space. Let (v^i), $1 \leq i,j,\ldots \leq n$, (summation convention) be a basis of V. Let (x_i) be the dual basis of V^d, i.e.,

$$x_i(v^j) = \delta_i^j \ .$$

Let $\mathscr{F}$ be the real-valued <u>polynomial</u> functions on V. The usual addition and multiplication define $\mathscr{F}$ as a <u>commutative associative algebra</u>.

A <u>vector field</u> (in the sense of classical tensor analysis) can be identified with a <u>derivation</u> of $\mathscr{F}$, i.e., a R-linear mapping A: $\mathscr{F} \to \mathscr{F}$ such that

$$A(f_1 f_2) = A(f_1)f_2 + f_1 A(f_2)$$

for $f_1, f_2 \in \mathscr{F}$.

Each $f \in \mathcal{F}$ can be written as a (polynomial) function $f(x^1,\ldots,x^n)$. Each such A can then be written as:

$$A(f) = A^i \frac{\partial f}{\partial x^i} \quad . \tag{6.1}$$

Now, $\mathcal{F}$ can be identified with the space of <u>symmetric</u>, multilinear maps on V, i.e., with the "boson" analogue of the Grassman algebra $\mathcal{G}(V)$ utilized in the preceding section. Thus, if

$$f: V \times \cdots \times V \to R$$

is such a symmetric multilinear map, (6.1) can be written as follows:

$$A(f) = A^i(v^i \lrcorner f) \quad , \tag{6.2}$$

The "product" used in (6.2) (the analogue of the exterior product $\wedge$ used in Section 5) is the <u>symmetric tensor product</u>.

Now, the set of vector fields of type (6.1) forms a Lie algebra under commutation. (The resulting Lie algebra operation is called the <u>Jacobi bracket</u> in differential geometry. We thus see that there is a natural and elegant analogy between the Lie algebra of vector fields and the Graded Lie Algebra $\mathcal{G}(V) \otimes V$ defined in Section 5.

This analogy has at least two consequences. First, it suggests how to develop a "fermion tensor analysis" by analogy with the usual kind--namely, consider <u>modules</u> of $\mathcal{G}(V)$, and linear representations of the GL $\to \mathcal{G}(V) \otimes V$ by means of

"Lie derivative" operations in these modules. Second, it suggests what we should mean by a "geometric" interpretation of a given graded Lie algebra--look for a homomorphism (in the graded Lie algebra sense) between it and the graded Lie algebra $\mathcal{G}(V) \otimes V$ for some vector space V.

Finally, notice that this algebraic interpretation of standard differential geometric ideas carries over to <u>infinite dimensional</u> vector spaces V. (Of course, the technical details cannot be handled quite so naively--but much will carry over.) In particular, this gives us a way to think of "differential geometry" for physical systems <u>with an infinite number of degrees of freedom</u>, e.g., field theories.

7. THE BELTRAMI BRACKET ON A RIEMANNIAN MANIFOLD

In classical mechanics, Lie algebras appear naturally via the <u>Poisson bracket operation</u>. The geometric structure associated is a <u>symplectic manifold</u>. This operation also plays a role in quantum mechanics, since at least some of the corresponding classical and quantum observables satisfy the same commutation and Poisson bracket relations.

In quantum mechanics, "observables" are Hermitian operators. They also admit a "commutative" operation, anticommutator. It is the combination of these two operations which leads to the notion of "graded Lie algebra".

In differential geometric terms, the "Riemannian metrics" play the same role in generating a symmetric first order differential operator on functions that the "symplectic structures" play in generating the Poisson bracket. This symmetric operation (that I shall call the Beltrami bracket) is a good candidate for providing a suitable geometric interpretation of "graded Lie algebras", and certain physical aspects of the theory of "fermions". It is not without interest that it also appears naturally in the geometric study of electric circuits [20,21,22]

For notation for differential geometry on manifolds, see [15], supplemented by the standard texts [18,19]. For the moment, we shall be working with finite dimensional manifolds, although there is strong physical interest in extending the ideas to the infinite dimensional case.

Let M be a manifold. $F(M)$ denotes the C^∞, real-valued function on M. $V(M)$ denotes the vector fields on M, which is an $F(M)$-module. Each $A \in V(M)$ is a derivative $f \to A(f)$ on $F(M)$. A Riemannian metric is an $F(M)$-bilinear, symmetric map

$$V(M) \times V(M) \to F(M)$$

which is non-degenerate. We denote it by $(A,B) \to <A,B>$. Suppose such a metric is given on M.

Let $f \in F(M)$. It determines a vector field $\mathrm{grad}(f)$, called the gradient, and defined as follows:

$$<\mathrm{grad}\ f, A> = A(f) \ . \tag{7.1}$$

The <u>Beltrami bracket</u> is a differential operator

$$F(M) \times F(M) \to F(M) ,$$

$$(f_1, f_2) \to <f_1, f_2>$$

defined as follows:

$$<f_1, f_2> = \text{grad } f_1(f_2) \qquad (7.2)$$

$$= <\text{grad } f_1, \text{grad } f_2> .$$

The symmetry of the matrix guarantees that

$$<f_1, f_2> = <f_2, f_1> .$$

A vector field A on M is said to be a <u>Killing vector field</u> or <u>infinitesimal isometry</u> if the Lie derivative of the metric tensor with respect to A is zero. This means that:

$$A<B,C> = <[A,B],C> + <B,[A,C]> \qquad (7.3)$$

$$\text{for } A,B,C \in V(M) .$$

The set of all Killing vector fields forms a Lie algebra. (It is finite dimensional, if M is a finite dimensional manifold.) If A is a Killing vector field, we readily derive the following relation between it and the Beltrami bracket:

$$A(<f_1, f_2>) = <A(f_1), f_2> + <f_1, A(f_2)> \qquad (7.4)$$

This relation is suggestive of the Jacobi bracket relation for Graded Lie Algebras. Let us now return to the purely algebraic situation in order to investigate this point.

8. Z_2-GRADED LIE ALGEBRAS

We would like to have an abstract algebraic structure which has both a commutative and anti-commutative bracket, but <u>not</u> a grading by integers. The notion of "Z_2-graded Lie algebra" is the appropriate one. To motivate its introduction, let us begin with a Graded Lie Algebra

$$GL = GL^0 \oplus GL^1 \oplus \cdots$$

as before. Recall that the bracket $[\ ,\ ]$ on it then satisfies the following relations:

$$[A,B] \subset GK^{n+m}$$

$$[A,B] = -(-1)^{nm}[B,A]$$

$$[A,[B,C]] = [[A,B],C] \oplus (-1)^{nm}[B,[A,C]]$$

for $A \in GL^n$, $B \in GL^m$, $C \in GL$.

Now, set:

$$GL^+ = GL^0 + GL^2 + GL^4 + \cdots$$

$$GL^- = GL^1 + GL^3 + \cdots$$

In words, GL^+ consists of the elements of even grading, GL^- the elements of odd grading. Notice that the following rules hold:

$$GL = GL^+ + GL^- \qquad (8.1)$$

$$GL^+ \cap GL^- = (0) \qquad (8.2)$$

$$[GL^+, GL^\pm] \subset GL^\pm \qquad (8.3)$$

$$[GL^-, GL^-] \subset GL^+ \qquad (8.4)$$

$$[A,B] = -[B,A] \qquad (8.5)$$
$$\text{if } A \in GL^+, \ B \in GL$$

$$[A,B] = -[B,A] \qquad (8.6)$$
$$\text{if } A,B \in GL^-$$

Definition. A Z_2-graded Lie algebra is defined as a vector space, with an algebra product, satisfying axioms (8.1)-(8.6).

Remark. Thus, the axioms for a Z_2-graded Lie algebra are a "weakening" of those for an integer-graded Lie algebra. Notice that there are similarities to the idea of "symmetric Lie subalgebra of a Lie algebra" [26].

With these algebraic concepts in hand, we can return to differential geometry.

9. Z_2-GRADED LIE ALGEBRAS WHICH ACT ON RIEMANNIAN-SYMPLECTIC MANIFOLDS

Let M be a manifold. M is said to have a <u>Riemannian-symplectic structure</u> if it carries a pair $(\omega, <,>)$ consisting of a closed non-degenerate two-form ω and a Riemannian metric $<,>$. As we have seen in Section 7, the Riemannian metric $<,>$ defines a symmetric product on $F(M)$ called the <u>Beltrami bracket</u>. Similarly, the symplectic structure ω defines a skew-symmetric bracket $\{ , \}$ on $F(M)$ called <u>Poisson bracket</u>. To define it, given $f \in F(M)$, let A be the vector field on M such that

$$df = A \lrcorner \omega \tag{9.1}$$

For $f, g \in F(M)$, set:

$$\{f, g\} = -A_f(g) \quad . \tag{9.2}$$

An obvious idea is then to try to "realize" a Z_2-graded Lie algebra so that the symmetric part of the bracket becomes the Beltrami bracket and the skew-symmetric part becomes Poisson bracket. Here is how to formalize this:

<u>Definition</u>. Let $GL = GL^+ \oplus GL^-$ be a Z_2-graded Lie algebra. A <u>realization</u> of GL on the symplectic-Riemannian manifold M is defined by an R-linear map

$$\phi: GL \to F(M)$$

satisfying the following conditions:

$\phi: GL^+ \to F(M)$ is a Lie algebra homomorphism
with respect to the given bracket on GL (9.3)
and the Poisson bracket on F(M).

$$\phi([\alpha,\beta]) = \{\phi(\alpha),\phi(\beta)\} \tag{9.4}$$
for $\alpha \in GL^+$, $\beta \in GL^-$

$$\phi([\alpha,\beta]) = <\phi(\alpha),\phi(\beta)> \tag{9.5}$$
for $\alpha,\beta \in GL^-$.

For each $\alpha \in GL^+$, the vector field A_α is such that

$$A_\alpha \lrcorner \omega = d\phi(\alpha)$$

is a Killing vector field.

Remark. This need to introduce <u>both</u> a symmetric and a skew-symmetric "metric" tensor in order to accommodate Z_2-graded Lie algebras is somewhat reminiscent (physically) of one of Einstein's approaches to a unified gravitational-electrical field theory, involving a "non-symmetric" metric tensor. In this connection, it is interesting to note Zumino's remarks in [12] that his approach is reminiscent of the Kaluza approach to a unified field theory!

10. KÄHLER MANIFOLDS

Having seen the need for manifolds carrying both symplectic and Riemannian structures having certain properties, one might naturally ask whether there are any such spaces. In fact, one class is well known to differential geometers-- the Kähler manifolds.

To define manifold M as a <u>Kähler manifold</u> one must be given the following data:

a) A Riemannian metric $<,>$. Denote the associated tension-free affine connection [15] by ∇.

b) An almost-complex structure, defined by an $F(M)$-linear map $J: V(M) \to V(M)$ such that $J^2 = -1$.

The covariant derivative of J with respect to ∇ is zero, and the following condition holds:

$$<JA,B> + <A,JB> = 0 \qquad (10.1)$$

for $A, B \in V(M)$.

Suppose such a Kähler manifold is given. One proves readily that the almost complex structure is integrable, i.e., M is a complex manifold in such a way that--in local complex analytic coordinates-- J is just multiplication by $\sqrt{-1}$. Set:

$$\omega(A,B) = <JA,B> \qquad (10.2)$$

for $A, B \in V(M)$.

It is a consequence of (10.1) that it is a non-degenerate, skew-symmetric form on vector fields, i.e., a two-differential form. Its covariant derivative with respect to ∇ is zero, hence, since ∇ is torsion-free,

$$d\omega = 0 \ .$$

Thus, ω defines M as a symplectic manifold as well.

For the rest of this chapter, let M have a fixed Kähler structure.

A vector field A on M is said to be an <u>infinitesimal Kähler isometry</u> or a <u>Kähler-Killing field</u> if the Lie derivative with respect to A of <u>both</u> the metric terms and the J-tensor is zero. It follows that the Lie derivative with respect to A of the symplectic form ω is zero as well. This implies that (at least locally) there is a function $f \in F(M)$ such that

$$A \lrcorner \omega = df \qquad (10.3)$$

Conversely, suppose that f is a function on M, and that A is a vector field that is <u>defined</u> by relation (10.3). Let us examine its relation to the gradient vector field grad f. (10.3) means that, for $B \in V(M)$,

$$B(f) = df(B) = \langle B, \text{grad } f \rangle = (A \lrcorner \omega)(B)$$

$$= \omega(A,B) = \langle JA, B \rangle \ .$$

Hence,

$$JA = \text{grad } f \ . \qquad (10.4)$$

Definition. Let $GL = GL^+ \oplus GL^-$ be a Z_2-graded Lie algebra. A realization of GL on the Kähler manifold M is defined as an R-linear map

$$\phi: GL \to F(M)$$

satisfying (9.3)-(9.5), and such that, in addition:

For $\alpha \in GL^+$, A_α, the vector field such that $A_\alpha \lrcorner \omega = d\phi(\alpha)$, is a Kähler-Killing vector field.

11. THE HERMITIAN-SKEW HERMITIAN Z_2-GRADED LIE ALGEBRA REALIZED ON PROJECTIVE SPACE

Let V be a complex vector space. It will be assumed to be **finite dimensional**, although, since we work in a coordinate free way, everything will carry over to the infinite dimensional situation--in particular, the space encountered as state spaces in quantum mechanics. Let PS(V) denote the **projective space** associated with V, i.e., a "point" of PS(V) is a one-dimensional linear complex subspace of V. Alternately PS(V) may be thought of as the orbit space of the action of the multiplicative group of non-zero complex numbers in V, i.e., $(\lambda, v) \to \lambda v$, for $\lambda \in \mathbb{C} - (0)$, $v \in V$. For $v \in V - (0)$, let (v) denote the one-dimensional linear subspace it generates, as an element of $PS(V)$. Then, the tangent space to the manifold $PS(V)$ at (v), denoted by $PS(V)_{(v)}$ is isomorphic naturally to the vector space

$$V/(v) \quad .$$

It is a complex vector space, and the complex structure is defined on it by means of J = multiplication by $\sqrt{-1}$.

In order to make $PS(V)$ into a Kähler manifold, introduce a Hermitian symmetric, complex-valued positive-definite bilinear form

$$(v_1, v_2) \to <v_1/v_2>$$

on V. With its aid, we can identify $V/(v)$ with the orthogonal complement of (v) in V — denoted by $(v)^\perp$ — hence, introduce a Hermitian symmetric, positive definite form on each tangent space $PS_{(v)}$. It is readily seen that this makes $PS(V)$ into a Kähler manifold.

Given a complex linear map

$$A: V \to V \quad ,$$

let A^* denote its adjoint with respect to the Hermitian form $</>$, i.e.,

$$<Av_1/v_2> = <v_1/A^*v_2> \quad .$$

A is said to be <u>Hermitian</u> if $A^* = A$, <u>skew-Hermitian</u> if $A^* = -A$. One can then define a Z_2-graded Lie algebra $GL = GL^+ \oplus GL^-$ as follows:

GL^+ = skew-Hermitian operators. Bracket is commutator

GL^- = Hermitian operators. Bracket is $\sqrt{-1}$ times anti-commutator.

This is, of course, the prototypical example of a graded Lie algebra as far as quantum mechanics is concerned.

To see the connection with Section 10, i.e., to define a realization of GL in the Kahler manifold PS(V), proceed as follows:

For $A \in GL^+$, $(v) \in PS(V)$,

$$\phi(A)((v)) = i<Av/v>/<v,v>$$

For $A \in GL^-$,

$$\phi(A)((v)) = \frac{<Av/v>}{<v/v>}$$

It is readily verified that all the conditions are satisfied.

12. GRADED LIE ALGEBRAS FOR CLASSICAL MECHANICAL PROBLEMS

Adopt the now-standard differential geometric setting for classical mechanics [15,23,24], i.e., Q is a manifold configuration space and $M = T^d(Q)$ is the cotangent bundle to M. It has a symplectic structure ω, which enables one to define a Poisson bracket operation on F(M) in the usual way. A function $h \in F(M)$, called the <u>Hamiltonian</u>, is defined so that, with

$$A_h \lrcorner \omega = dh \quad ,$$

A_h is a vector field whose orbit curves are the time-trajectories of the mechanical system.

In many problems of mechanics, a set $\underset{\sim}{S}$ of functions on M will be given which is closed under Poisson bracket, i.e., form a Lie subalgebra of $F(M)$, and which either satisfies

$$\{h, \underset{\sim}{S}\} = 0 \quad ,$$

or some other more general sort of commutation relation with h. (I am thinking of the material that is generally called the theory of <u>spectrum generating algebras</u> [25] in classical mechanics and the quantum mechanical problems, e.g., harmonic oscillatory hydrogen atom, rigid rotation, which are closely linked to classical mechanics.)

It is a natural idea to look for systems for which there is a symmetric bracket available, so that $\underset{\sim}{S}$ can become a Graded Lie Algebra. This will give an additional structure to the mechanical system which might be useful for various purposes, e.g., in quantization.

Let us look at this for the simplest sort of general system, the Harmonic Oscillators. Here,

$$Q = R^n \quad .$$

Choose indices and the summation convention as follows:

$$1 \le i,j,k \le n \quad .$$

Let (q^i) be linear coordinates on R^n. The natural coordinates on $M = T^d(q) = Q \times Q^d$ are

$$(q^i, p_i) \quad ,$$

where (p_j) are essentially the dual coordinates to the (q^i). The symplectic form on M is

$$\omega = dp_i \wedge dq^i \quad .$$

Suppose that:

$$h = \tfrac{1}{2}((q)^2 + (p)^2)$$

with

$$(q)^2 = q^1 q^1 + \cdots, \quad (p)^2 = p_1 p_1 + \cdots$$

Hamilton's equations are then

$$\frac{dq}{dt} = p$$

$$\frac{dp}{dt} = -q \quad ,$$

i.e., the solutions are orbit spaces of

$$A_h = p_i \frac{\partial}{\partial q^i} - q^i \frac{\partial}{\partial p_i}$$

so that this is recognized as the Hamiltonian of the harmonic oscillator.

Let $\underset{\sim}{S}$ consist of the functions of the quadratic type:

$$f(p,q) = A^{ij}p_i p_j + B^i_j p_i q^j + C_{ij}q^i q^j \qquad (12.1)$$

such that

$$\{f,h\} = 0 \quad . \qquad (12.2)$$

The set of <u>all</u> functions of the form (12.1) forms (under Poisson bracket) the Lie algebra of the symplectic Lie group $Sp(n,R)$. (See [26] for Lie group notation.) It is well known that $\underset{\sim}{S}$ forms the Lie algebra of the <u>maximal compact subgroup</u> of $Sp(n,R)$, which is the unitary group $U(n)$. (Indeed, this is one basic reason that the unitary group plays a role in nuclear physics, independently of its possible role in elementary particle physics.) We would now like to define a Kähler motion on M, and a set of functions whose Beltrami bracket gives elements of $\underset{\sim}{S}$. Since everything else is linear, it is natural to expect that the metric will also be linear. Hence, let us look at the question from a point of view of "pure" linear algebra. (This also has the benefit of suggesting infinite dimensional generalizations.)

Change the notation accordingly. Let V be an even dimensional real vector space, with a non-degenerate skew-symmetric bilinear form

$$\omega: V \times V \to R \quad .$$

(In the situation described above, V is the direct sum of Q and its dual vector space.) h is a positive definite symmetric bilinear form $V \times V \to R$.

If f is any symmetric bilinear form: $V \times V \to R$, it can be written as follows:

$$f(v_1, v_2) = \omega(A_f v_1, v_2) \quad . \tag{12.3}$$

for $v_1, v_2 \in V$, with A_f a linear map $V \to V$.

Then,

$$\omega(A_f v_1, v_2) = f(v_1, v_2) = f(v_2, v_1)$$

$$= \omega(A_f v_2, v_1)$$

or

$$\omega(A_f v_1, v_2) + \omega(v_1, A_f v_2) = 0 \tag{12.4}$$

(12.4) says that A_f generates a one-parameter group of automorphisms of the skew-symmetric form ω. In particular, this assignment

$$f \to A_f$$

is the coordinate-free version of the relation (12.1), establishing an isomorphism between the Lie algebra of $Sp(n,R)$ and the space of $2n \times 2n$ symmetric matrices.

Let $h: V \times V \to R$ be a fixed positive definite symmetric bilinear form, which physically represents the <u>energy</u>. Let $\underset{\sim}{S}$ denote the set of symmetric bilinear forms f such that

$$\{A_h, A_f\} = 0$$

For $f \in \underset{\sim}{S}$, let

$$B_h: V \to V$$

be the linear map such that

$$f(v_1, v_2) = h(B_f(v_1), v_2) \quad . \tag{12.5}$$

For $f, g \in \underset{\sim}{S}$, set:

$$[f,g]_-(v_1, v_2) = h((B_f B_g + B_g B_f)(v_1), v_2)$$

$$= h(B_g(v_1), B_f(v_2)) + h(B_f(v_1), B_g(v_2))$$

$[f,g]_-$ is a symmetric bilinear form. Let $J: V \to V$ be the map such that

$$h(v_1, Jv_2) = \omega(Jv_1, v_2) \quad ,$$

$$J^2 = -1 \quad .$$

Then,

$$A_{[f,g]_-} = J(B_f B_g + B_g B_f) \quad . \tag{12.6}$$

We must still use the condition that $f \in \underset{\sim}{S}$. It is clear that A_f commutes with J. Now,

$$h(B_f v_1, v_2) = \omega(A_f v_1, v_2) = \omega(JB_f v_1, v_2) \quad ,$$

i.e.,

$$A_f = JB_f \quad .$$

Hence, the condition that $f \in \underset{\sim}{S}$ is that J commutes with B_f. This finally proves that:

$$[\underset{\sim}{S},\underset{\sim}{S}]_- \subset \underset{\sim}{S} \tag{12.7}$$

We can then define a Graded Lie Algebra $GL = GL^+ \oplus GL^-$, with both pieces isomorphic as <u>vector spaces</u> to $\underset{\sim}{S}$, with the bracket in each defined essentially as the commutator and anti-commutator.

<u>Remark</u>. One property of this construction that is probably important for generalization later on is that Beltrami bracket is <u>dependent on the energy</u> of the system, whereas the Poisson bracket is not.

13. LINEAR HEISENBERG-FERMI STRUCTURES ON A VECTOR SPACE AND THEIR FOCK REPRESENTATIONS

In the last section we have seen that harmonic oscillators define the following algebraic structure:

> A real vector space V, with a pair (ω,θ) of bilinear forms $V \times V \to R$ with ω skew-symmetric, θ symmetric.

These structures also play a role in quantum field theory, in which case, V is infinite dimensional. Hence, it is worthwhile giving it a name--we will call such a triple (V,ω,θ) a linear Heisenberg-Fermi structure.

If one introduces an abstract structure of this sort, one is accustomed to then describing its <u>representations</u>. I intend to come back to this more systematically at another point; here, I want to just say enough to indicate the relation to the general theory of graded Lie algebras and current algebras.

Let (V,ω,θ) be such a structure. In order to "represent" it (in ways that are of interest for quantum mechanics) introduce another abstract structure.

<u>Definition</u>. A <u>linear Hermitian structure</u> is a pair consisting of a complex vector space H (whose elements are denoted by ψ), and an R-bilinear map

$$H \times H \to \mathbb{C} ,$$

$$(\psi_1,\psi_2) \to <\psi_1/\psi_2> ,$$

such that:

$$<\psi_1/\psi_2>^* = <\psi_2/\psi_1>$$

$$<\psi_1/c\psi_2> = c<\psi_1/\psi_2>$$

$$\text{for } c \in \mathbb{C} ; \quad \psi_1,\psi_2 \in H .$$

If $A: H \to H$ is a complex linear map, its <u>adjoint</u> is a complex linear map $A^*: H \to H$ such that

$$\langle A\psi_1/\psi_2 \rangle = \langle \psi_1/A^*\psi_2 \rangle \quad .$$

Definition. Let (V,ω,θ) be a linear Heisenberg-Fermi structure. A <u>Fock-Bose</u> or <u>Fock-Fermi</u> representation is defined as an R-linear map $cr: V \to$ (linear maps on a Hermitian space H) such that

$$cr(v_1)cr(v_2) \mp cr(v_2)cr(v_1) = 0 \qquad (13.1)$$

$$cr(v_1)^* cr(v_2) \mp cr(v_2)cr(v_1)^*$$
$$= (\theta(v_1,v_2) + i\omega(v_1,v_2))1 \qquad (13.2)$$

for $v_1, v_2 \in V$.

<u>Remark.</u> In (13.1) and (13.2), "-" means Bose, "+" means Fermi, commutation relations. "1" means the identity map $H \to H$. "cr" stands for "creation operator", since it is in these terms that representations like this are constructed in physics. "$cr(v)^*$" is usually called an <u>annihilation operator</u>.

Let us suppose such a representation is given. To make it useful for quantum mechanical purposes, Hermitian (or skew-Hermitian) operators should be constructed. This will lead us to Lie algebras and graded Lie algebras of the "Heisenberg" type.

For $v \in V$, set:

$$\rho(v) = cr(v) + \lambda\, cr(v)^* \qquad (13.3)$$

(λ is a real number, which ultimately will be chosen as ± 1.)
Then,

$$\rho(v_1)\rho(v_2) = (cr(v_1) + \lambda\, cr(v_1)^*)(cr(v_2) + \lambda\, cr(v_2)^*)$$

$$= cr(v_1)cr(v_2) + \lambda(cr(v_1)^* cr(v_2)$$
$$+ cr(v_1)cr(v_2)^*) + \lambda^2 cr(v_1)^* cr(v_2)^*$$

Hence,

$$\rho(v_1)\rho(v_2) \mp \rho(v_2)\rho(v_1) = cr(v_1)cr(v_2) \mp cr(v_2)cr(v_1)$$
$$+ \lambda^2(cr(v_1)^* cr(v_2)^* \mp cr(v_2)^* cr(v_1)^*)$$
$$+ \lambda(cr(v_1)^* cr(v_2) \mp cr(v_2)^* cr(v_1)$$
$$+ cr(v_1)cr(v_2)^* \mp cr(v_2)cr(v_1)^*)$$

$$= 0 + \lambda^2 \cdot 0 + \lambda(\theta(v_1,\theta_2) + i\omega(v_1,v_2)$$
$$\mp (cr(v_2)^* cr(v_1) \mp cr(v_1)cr(v_2)^*))$$

$$= \lambda(\theta(v_1,v_2) + i\omega(v_1,v_2)$$
$$\mp (\theta(v_2,v_1) + i\omega(v_2,v_1))$$

$$= \lambda(\theta(v_1,v_2)(1 \mp 1) + i\omega(v_1,v_2)(1 \pm 1)$$
$$\qquad (13.4)$$

This formula presents the Bose and Fermi commutation relations in a unified form. To obtain the usual formulas, proceed as follows.

Bose. Take $\lambda = -1$, so that $\rho(v)$ is skew-Hermitian:

$$\rho(v_1)\rho(v_2) - \rho(v_2)\rho(v_1) = -2i\omega(v_1, v_2) \qquad (13.5)$$

Fermi. Take $\lambda = +1$, so that $\rho(v)$ is Hermitian:

$$\rho(v_1)\rho(v_2) + \rho(v_2)\rho(v_1) = 2\theta(v_1, v_2) \qquad (13.6)$$

These formulas can be most readily interpreted in terms of graded Lie algebras. In the Bose case, let $\underset{\sim}{G}$ be a Lie algebra constructed as follows:

As a vector space, $\underset{\sim}{G}$ is a direct sum of V and R.

$$[V, R] = 0$$
$$[v_1, v_2] = -2\omega(v_1, v_2) \qquad (13.7)$$

$\underset{\sim}{G}$ is called the Heisenberg Lie algebra.

In the Fermi case, construct a Z_2-graded Lie algebra GL as follows:

$$GL^+ = R ,$$
$$GL^- = V ,$$

$$[GL^+, GL^-] = 0$$

$$[v_1, v_2] = 2\theta(v_1, v_2)$$

(13.8)

This is called the <u>Heisenberg-Fermi graded Lie algebra</u>.

In both cases, we see that the formulas given above construct representations. They appear in many physics contexts.

Bibliography

1. R. Hermann, "A Geometric Formula for Current Algebra Commutation Relations", Phys. Rev. 177 (1969), p. 2449.

2. R. Hermann, "Quantum Field Theories with Degenerate Lagrangians", Phys. Rev. 177 (1969), p. 2453.

3. R. Hermann, Fourier Analysis on Groups and Partial Wave Analysis, W.A. Benjamin, New York, 1969.

4. R. Hermann, "Algebraic and Geometric Structures in Current Algebra Theory", ONR Technical Report No. 1, Institute for Advanced Study, Princeton, New Jersey, 1970.

5. R. Hermann, "Current Algebra, Sugawara Model and Differential Geometry", J. Math Phys. 11 (1970), pp. 1825-1829.

6. R. Hermann, "Infinite Dimensional Lie Algebra and Current Algebra", Proc. 1969 Battele-Seattle Recontres on Math. Physics, Springer Verlag, Berlin, 1970, pp. 312-337.

7. R. Hermann, Lie Algebras and Quantum Mechanics, W.A. Benjamin, New York, 1970, 400pp.

8. R. Hermann, "Currents in Classical Field Theories, J. Math. Phys. 13 (1972), p. 97.

9. R. Hermann, Topics in the Mathematics of Quantum Mechanics Vol. VI of Interdisciplinary Mathematics, Math Sci Press, Brookline, Mass., 1973.

10. J. Wess and B. Zumino, Nucl. Phys. B70, (1974), p. 39.

11. L. Corwin, Y. Ne'eman and S. Sternberg, Reviews of Modern Physics 47 (1975).

12. R. Arnowitt and P. Nath (eds.), Gauge Theories and Modern Field Theory, MIT Press, 1976.

13. W. Greub, Multilinear Algebra, Springer-Verlag, Berlin, 1967.

14. R. Hermann, Linear and Tensor Algebra, Vol. II of Interdisciplinary Mathematics, Math Sci Press, Brookline, Mass., 1973.

15. R. Hermann, Differential Geometry and the Calculus of Variations, Academic Press, New York, 1969.

16. B. Kostant, *Graded Manifolds, Graded Lie Theory and Prequantization*, MIT Preprint, 1976.

17. S. Sternberg, *Lectures on Differential Geometry*, Prentice-Hall, 1965.

18. R. Bishop and S. Goldberg, *Tensor Analysis on Manifolds*, Macmillan, New York, 1968.

19. W. Boothby, *An Introduction to Differential Manifolds and Riemannian Geometry*, Academic Press, New York, 1975.

20. A.K. Brayton and J.K. Moser, *Quart. Appl. Math.* **22** (1964), pp. 1-33.

21. S. Smale, "On the Mathematical Foundations of Electric Circuit Theory", *J. Diff. Geom.* **7** (1972), pp. 193-210.

22. R. Brockett and R.A. Skoog, "A New Perturbation Theory for the Synthesis of Nonlinear Networks", SIAM-AMS Proceedings, Vol. III, American Math. Soc., Providence Rhode Island, 1971.

23. R. Abraham and J. Marsden, *Foundations of Mechanics*, W.A. Benjamin, New York, 1966.

24. J.M. Souriau, *Structures des Systèmes Dynamiques*, Durod, Paris, 1970.

25. R. Hermann, "Spectrum-Generating Algebras in Classical Mechanics", I and II, *J. Math. Phys.* **13** (1972), pp. 833, 878.

26. S. Helgason, *Differential Geometry and Symmetric Spaces*, Academic Press, New York, 1962.

V.

THE DIFFERENTIAL-GEOMETRIC FOUNDATIONS OF "SUPERSPACE" AND "SUPERSYMMETRY" THEORIES

PART II: THE QUADRATIC DIFFERENTIAL CO-FORMS ON THE DUAL SPACE OF A GRADED LIE ALGEBRA

Abstract

After presenting some general definitions concerning bracket operations for functions on manifolds, a bracket structure is exhibited for the dual space of a graded Lie algebra. This leads to generalizations of the classical spinning-particle equations.

1. INTRODUCTION

There are many equivalent approaches to the differential geometry of a classical spinning particle. Berezin and Marinov have recently presented [1] a description in terms of "fermion" differential geometry. In [2], Chapter 14, I have presented a description in terms of what I called a <u>cosymplectic structure</u> on the dual space to an ordinary Lie algebra. While attempting to understand their interrelation, I found it useful to extend the formalism of [2] from "Lie algebras" to "Z_2-graded Lie algebras". In Part I of this work I also remarked that "graded Lie algebras" could be realized in terms of the usual differential manifolds if we introduced a pair consisting of a symmetric and skew-symmetric quadratic differential form. In fact, the main point that this paper makes, is that a Graded Lie Algebra defines such an object in its dual space in a natural way, and that this is a very natural and interesting mathematical structure for the study of many physics situations.

2. CO-RIEMANNIAN AND CO-SYMPLECTIC STRUCTURE. QUADRATIC DIFFERENTIAL CO-FORMS

In [2] and [3] it was remarked that for certain physical (and mathematical) purposes, it was convenient to introduce and study differential geometric objects which are dual to the usual notions of "Riemannian" and "symplectic" structures.

(In terms of tensor analysis, they are "contravariant tensors" rather than "covariant" ones.) Let us now recall their definition.

Let M be a differentiable manifold in the usual sense. Let $F(M)$ denote its C^∞ real-valued functions. Let $V(M)$ denote the C^∞ vector fields, considered as an $F(M)$-module. $F^1(M)$ --the one-<u>differential forms</u>--are the dual $F(M)$-module to $V(M)$.

<u>Definition</u>. A <u>quadratic differential co-form on</u> M is an $F(M)$-bilinear mapping

$$q: F^1(M) \times F^1(M) \to F(M) \quad .$$

It is said to be <u>co-Riemannian</u> if q is <u>symmetric</u>, <u>co-symplectic</u> if q is skew-symmetric.

If q is non-degenerate, i.e.,

$$q(\theta, F^1(M)) = 0 \Rightarrow \theta = 0$$

$$\text{for } \theta \in F^1(M) \quad ,$$

then q dualizes to define a quadratic differential form in the usual sense, i.e., an $F(M)$-bilinear map

$$V(M) \times V(M) \to F(M) \quad .$$

Such a q defines a first order differential operation on functions, called the <u>Bracket</u> (<u>Poisson</u>, if q is skew-symmetric, <u>Beltrami</u>, if q is symmetric). It is an R-bilinear map

$$F(M) \times F(M) \to F(M) \quad,$$

denoted by

$$(f_1, f_2) \to \{f_1, f_2\} \quad,$$

defined as follows:

$$\{f_1, f_2\} = q(df_1, df_2) \quad.$$

The algebraic rules satisfies by this bracket are readily derived from the formula. For example,

$$\{f_1 f_2, f_3\} = q(d(f_1, f_2), df f_3)$$

$$= f_1 \{f_2, f_3\} + f_2 \{f_1, f_3\}$$

$$\{f_1, f_2 f_3\} = q(df_1, d(f_2, f_3))$$

$$= f_3 \{f_1, f_2\} + f_2 \{f_1, f_3\} \quad.$$

Such a q can also be regarded as a "tensor field" in the sense of classical tensor analysis. Let (x^i), $1 \leq i, j \leq m$, be a coordinate system for M. Then,

$$q^{ij} = q(dx^i, dx^j) \equiv \{x^i, x^j\}$$

are the <u>components of the tensor field in this coordinate system</u>. It is readily verified that they transform like a <u>rank two-contravariant tensor</u> on change of coordinate system.

3. MORPHISMS AND PRODUCTS OF QUADRATIC DIFFERENTIAL CO-FORMS

In modern categorical-functorial mathematics one learns that it is always worth while to "look for the morphisms". Such concerns also often have physical and geometric significance.

Let q and q' be quadratic differential co-forms on manifolds M and M', and let

$$\phi: M \to M'$$

be a smooth (i.e., C^∞) map between them. As usual in differential geometry, ϕ induces a "pull back" map

$$\phi^*: F(M') \to F(M)$$

$$\phi^*: F^1(M) \to F^1(M)$$

ϕ is said to be a <u>morphism</u> between q and q' if the following condition is satisfied:

$$\phi^*(q'(\theta_1', \theta_2')) = q(\phi^*(\theta_1'), \phi^*(\theta_2')) \qquad (3.1)$$

for $\theta_1', \theta_2' \in F^1(M')$.

For example, if q and q' are non-degenerate Riemannian co-metrics, then condition (3.1) amounts to saying that ϕ is a <u>Riemannian submersion</u> between the Riemannian metrics they determine. See [3].

Given (M,q), (M',q'), one can form their <u>products</u> (M'',q'').

$$M'' = M \times M' \equiv \text{the Cartesian product manifold}$$

For each $p'' = (p,p') \in M$, the tangent space $M''_{p''}$ is the direct sum $M_p \oplus M'_{p'}$ of the tangent spaces. Then, the value of q'' at p'' is the direct sum of the values of q, q' at p, p', i.e.,

$$q''(p'')(v_1 \oplus v'_1, v_2 \oplus v'_2) = q(v_1,v_2) + q'(v'_1,v'_2) \qquad (3.2)$$

It is readily seen that the Cartesian product maps

$$\phi: M'' \to M$$

$$\phi': M'' \to M'$$

are morphisms, in the sense defined above.

As in [4], one can now define "interactions" between (M,q), (M',q') as appropriately chosen submanifolds of $M \times M'$.

4. THE DYNAMICS DETERMINED BY A QUADRATIC DIFFERENTIAL CO-FORM AND A FUNCTION

Let (M,q) be a quadratic differential co-form. It defines a bracket operation

$$(f_1, f_2) \to \{f_1, f_2\} = q(df_1, df_2)$$

on functions. Let $h \in F(M)$. The <u>dynamical equation</u> is

$$\frac{\partial f}{\partial t} = \{h, f\} \quad . \qquad (4.1)$$

when q is chosen suitably, it can be either Liouville's equation of classical mechanics or the Brayton-Moser equations of motion of an electric circuit [5].

(4.1) is a first order partial differential equation on $M \times R$. Its "Cauchy characteristic" curves may be considered the "trajectories" of the motion. They can be defined as follows. The map

$$f \to A_h(f) = q(df, h)$$

if $F(M) \to F(M)$ satisfies the following:

$$A_h(f_1 f_2) = A_h(f_1) f_2 + f_1 A_h(f_2) \ .$$

It is then a vector field on M; the trajectories of (4.1) are the orbits of A_h, i.e., the curve $t \to p(t)$ such that

$$\frac{dp}{dt} = A_h(p(t)) \ . \qquad (4.2)$$

5. THE QUADRATIC DIFFERENTIAL CO-FORMS ON THE DUAL SPACE TO A GRADED LIE ALGEBRA

In this Part II when we speak of "graded Lie algebras" we mean "Z_2-graded Lie algebras". Thus, one, denoted by GL, is a vector space direct sum

$$GL = GL^+ \oplus GL^- \ ,$$

with:

$$[GL^+, GL^\pm] \subset GL^\pm \ , \qquad [GL^-, GL^-] \subset GL^+$$

[,] is symmetric on GL^-, skew-symmetric on GL^+.

Suppose, for simplicity, that GL is finite dimensional. Let M be its dual vector space, i.e., an element $p \in M$ is a linear map

$$GL \to R \quad .$$

Then, the tangent space M_p can be identified--as for any vector space--with M_p itself, i.e., with GL^d. Hence, its dual space M_p^d is GL itself. Hence, a bilinear form $M_p^d \times M_p^d \to R$ is identified with a bilinear form $GL \times GL \to R$.

Now, define a quadratic differential co-form q on M as follows:

The value of q at $p \in M \equiv GL^d$ is the following bilinear form

$$q(A,B) = p([A,B]) \qquad (5.1)$$

for $A, B \in GL \equiv (GL^d)^d \equiv M_p^d$

Now, GL itself is embedded in F(M), namely, as the <u>linear</u> real-valued functions, i.e.,

$$f_A(p) = p(A) \qquad (5.2)$$

for $A \in GL$, $p \in M \equiv GL^d$.

Hence,

$$df_A(p_0) = p_0(A)$$

for $A \in GL$, $p_0 \in M_p \equiv GL^d$.

$$\{f_A, f_B\}(p) = q(df_A, df_B)(p)$$

$$= p([A,B])$$

$$= f_{[A,B]}(p) \ .$$

Hence, we have proved:

$$\{f_A, f_V\} = f_{[A,B]} \qquad (5.3)$$
$$\text{for } A, B \in GL \ .$$

Thus, if h is a linear function on M defined by

$$h(p) = p(A) \ ,$$

where A is a fixed element of GL, then the equations of motion are

$$\frac{dB}{dt} = [A,B] \ , \qquad (5.4)$$

which is the natural thing from the group-theoretic point of view.

However, there are other functions in F(M) more complicated than the linear ones, namely, the <u>tensors</u> on GL. We now turn to their study.

6. TENSORIAL FUNCTIONS ON THE DUAL SPACE TO A GRADED LIE ALGEBRA

Let GL, $M \equiv GL^d$, $F(M)$ be as in Section 5. Let $T(GL)$ denote the (associative) tensor algebra of the vector space GL. Each

$$\tau \in T(GL)$$

then defines a real-valued function

$$f_\tau : M \to R$$

as follows: If $\tau = A_1 \otimes \cdots \otimes A_r$, then

$$f_\tau(p) = p(A_1) \cdots p(A_r) \qquad (6.1)$$

It is readily seen that this mapping

$$T(GL) \to F(M) \qquad (6.2)$$

is an algebra homomorphism.

The kernel is an ideal that contains all elements of the form

$$A \otimes B - B \otimes A \quad .$$

Hence, the map (6.2) passes to the quotient to define a map

$$S(GL) \to F(M) \quad , \qquad (6.3)$$

where $S(GL)$ denotes the <u>symmetric</u> tensor algebra based on the vector space GL. It is readily verified that the map of

S(GL) under the map (6.3) is closed under the $\{\ ,\ \}$-bracket; this provides S(GL) with a bracket structure. (In case GL is an ordinary Lie algebra, this bracket structure is that defined in [6].) For example, it plays a basic role in the study of the <u>rotating rigid body</u>.

We now turn to the main point of this paper--a description of a set of ordinary differential equations defined by a Graded Lie algebra which are a generalization of Euler's rigid rotating body equations.

7. THE EQUATIONS OF MOTION DETERMINED BY A QUADRATIC TENSOR

Let GL, M be as before. Let A^i, $1 \leq i,j \leq m$ be a basis for GL. Then,

$$[A^i, A^j] = c_k^{ij} A^k \qquad (7.1)$$

The (c_k^{ij}) are the <u>structure constants</u> of the graded Lie algebra. Let (α_i) be a dual basis of $GL^d \equiv M$, i.e.,

$$\alpha_i(A^j) = \delta_i^j \ .$$

Then, there are real-valued functions x^i on M such that:

$$p = x^i(p)\alpha_i \ ,$$

for $p \in M$,

i.e., the x^i are the <u>linear coordinates</u> on M. (They are also the functions f_{A^i} defined in the previous section.)

Let $\{\ ,\ \}$ be the bracket on $F(M)$ defined as explained in previous sections. Then,

$$\{x^i, x^j\} = c_k^{ij} x^k \quad . \tag{7.2}$$

Set

$$h = h_{ij} x^i x^j \quad . \tag{7.3}$$

For $f \in F(M)$,

$$\{h, f\} = h_{ij}(\{x^i, f\} x^j + x^i \{x^j, f\}) \quad .$$

Let A_h be the vector field on M such that

$$A_h(f) = h, f$$

Suppose

$$A_h = A^i \frac{\partial}{\partial x^i} \quad .$$

Then,

$$A^i = \{h, x^i\} = h_{jk}(\{x^j, x^i\} x^k + x^j \{x^k, x^i\})$$

$$= h_{jk}(c_k^{ji} x^\ell x^k + c_\ell^{ki} x^\ell x^j) \tag{7.4}$$

The dynamical equations are then just the ordinary (nonlinear) differential equations

$$\frac{dx^i}{dt} = A^i(x(t)) \quad . \tag{7.5}$$

Remark. We have not at all used the specific axiomatic properties of a graded Lie algebra--any algebra would have done as well!

Bibliography

1. F. Berezin and M. Marinov, "Particle Spin Dynamics as the Grassman Variant of Classical Mechanics", *Annals of Physics* **104** (1977), 336-362.

2. R. Hermann, *Gauge Fields and Cartan-Ehresmann Connections*, Part A (Interdisciplinary Mathematics, Vol. 10), Math Sci Press, Brookline, Mass., 1975.

3. R. Hermann, *Topics in General Relativity* (Interdisciplinary Mathematics, Vol. V), Math Sci Press, Brookline, Mass., 1975.

4. R. Hermann, *Geometric Structure of Systems-Control Theory and Physics*, Part A (Interdisciplinary Mathematics, Vol. IX), Math Sci Press, Brookline, Mass., 1975.

5. A. Brayton and J. Moser, *Quart. Appl. Math.* **22** (1964), pp. 1-33.

6. R. Hermann, "Spectrum Generating Algebra in Classical Mechanics", Parts I and II, *J. Math. Phys.* **13** (1972), pp. 833 and 878.

VI.

THE DIFFERENTIAL-GEOMETRIC FOUNDATIONS OF "SUPERSYMMETRY" AND "SUPERSPACE" THEORIES, III:

LAGRANGIAN FIELD THEORIES IN FERMION AND QUANTUM DIFFERENTIAL GEOMETRY

Abstract

A sketch is presented of a method of defining a "differential form algebra" based on arbitrary associative algebra. With its aid, the "Cartan form" can be defined for an arbitrary Lagrangian field theory, and a differential form calculus can be developed. As application, ideas of Berezin and Marinov--concerning a "Poisson bracket" for Fermion spaces and the resulting geometric interpretation of the Dirac equation--are efficiently developed.

1. Introduction

The physicists have developed a broad approach to the problem of constructing appropriate mathematical models for their ideas that they call "Lagrangian field theory". This material has both "classical" and "quantum" aspects, and (ultimately) involves Lie group theory and differential geometry as much as it does the "analysis" of traditional mathematical physics. In previous work [1-24] I have developed much of this material from the point of view of modern differential geometry, i.e., differential forms, differentiable manifolds, fiber spaces, etc.

When I was doing this work (which was more-or-less completed with Geometry, Physics and Systems [24]) I never really understood how "Fermions" fit into the picture. (I was not alone!) However, since then, the subject of "graded Lie algebras" (which the physicists call "supersymmetries") has come along and has turned out to be the appropriate mathematical framework needed to understand these objects. (The key papers in my own understanding were those of Zumino and Wess [25], [26], Zumino [27], Corwin, Ne'eman and Sternberg [28], Kostant [29], Berezin and Kac [30], Berezin [31], and Berezin and Marinov [32], although I am not familiar enough with the complete scholarly picture here to know if there is other relevant work. I suspect there is--for example, I heard, in 1972, of a derivation of Dirac equations by B. Sakita which had much of the flavor of this business, although I did not understand it at the time.)

The starting point for this paper is a brief note of mine [24] which sketches how differential forms may be constructed and how they may be applied in more general contexts than the traditional manifold theory. Having defined differential forms with appropriate properties, the route is open to generalizing the material previously worked out [1-23], in the "ordinary" manifold context. In fact, this process of generalization is surprisingly routine, hence in this paper I will only present a few sample ideas.

The point to these generalizations is that most of the concepts that seem to be exciting to elementary particle physicsts today (e.g., Yang-Mills-gauge fields, solitons, instantons, monopoles,...) are describable in terms of these "modern" differential geometric ideas. (For example, for "Yang-Mills", see [7], [11]. For "solitons", see [23].) I strongly suspect that with more work everything of interest can be described in this way, and much that is new, besides. Thus, we may soon appreciate that "modern" differential geometry is as closely linked to elementary particle physics as "Hilbert spaces" are linked to quantum mechanics. (The physicists do not realize this yet--they are so mathematically "pragmatic" that they have largely lost the talent and taste for creative mathematical speculation that played such an important role in the successes of the past.)

2. Differential Form Algebras

All algebras will have the reals or complex numbers as field of scalars.

Let $\mathcal{D}$ be an associative algebra. Denote the algebra operation by "$\wedge$". Suppose it has a unit element denoted by "1". Suppose that $\mathcal{D}$ has a subalgebra $\mathcal{A}$, which contains the unit. For notational convenience, we will denote the algebra operation on $\mathcal{A}$ by simple juxtaposition, i.e.,

$$a_1 \wedge a_2 = a_1 a_2$$
$$\text{for } a_1, a_2 \in \mathcal{A}.$$

Let d be a linear map $\mathcal{D} \to \mathcal{D}$ such that:

$$dd = 0 \tag{2.1}$$

$$d(a\omega) = da \wedge \omega + a d\omega$$
$$\text{for } a \in \mathcal{A}, \ \omega \in \mathcal{D}$$

Every element of $\mathcal{D}$ can be written as a sum of elements of the form

$$a da_1 \wedge \cdots \wedge da_r , \tag{2.2}$$

with $a, a_1, \ldots, a_r \in \mathcal{A}$, r arbitrary.

$$d(da \wedge \omega) = -da \wedge d\omega \tag{2.3}$$
$$\text{for } a \in \mathcal{A}, \ \omega \in \mathcal{D}$$

If these conditions are satisfied, we will call ($\mathcal{A}$, $\mathcal{D}$, d) a differential form algebra.

Example. Differential forms on a manifold.

Let M be a manifold, and let $\mathcal{A}$ be the algebra (under pointwise multiplication and addition) of C^∞, real-valued functions on M. Let $\mathcal{D}$ be the direct sum of differential forms [4] of all degrees, let ∧ be the usual exterior multiplication and let d be the exterior derivation. All these conditions are then satisfied. In addition, we have the following commutation relations:

$$a_1 a_2 = a_2 a_1$$

$$(da_1) a_2 = a_2 (da_1)$$

$$da_1 \wedge da_2 = -da_2 \wedge da_1$$

for $a_1, a_2 \in \mathcal{A}$

These formulas completely determine the commutation relations in $\mathcal{D}$.

3. <u>Boson and Fermion Differential Form Algebras</u>: "<u>Superspace</u>"

Let ($\mathcal{D}$, $\mathcal{A}$, ∧, d) be as in Section 2. It will be called a Boson differential form algebra if the following conditions are satisfied:

$\mathcal{A}$ is generated by elements $x^1,\ldots,x^n$ such that:

$$x^i x^j = x^j x^i \tag{3.1}$$

$$x^i dx^j = x^j dx^i \tag{3.2}$$

$$dx^i \wedge dx^j = -dx^j \wedge dx^i \tag{3.3}$$

for $1 \le i,j \le n$.

These are the only relations satisfied by the (x^i), i.e., $\mathcal{A}$ is the polynomial algebra in the elements (x^i).

It will be called a <u>Fermion differential form algebra</u> if $\mathcal{A}$ is generated by elements (y) $1 \le u,v \le m$ such that:

$$y^u y^v = -y^v y^u \tag{3.4}$$

$$y^u dy^v = -dy^v y^u \tag{3.5}$$

$$dy^u \wedge dy^v = dy^v \wedge dy^u \tag{3.6}$$

These are the only relations satisfied by the (y^u), i.e., $\mathcal{A}$ is the Grassman algebra in the variable (y^u).

Using these rules, we see that in the Boson case, each element of $\mathcal{D}$ can be written as sum of elements of the form

$$a_{j_1\cdots j_r}\, dx^{i_1} \wedge \cdots \wedge dx^{i_r} \quad,$$

with coefficients depending skew-symmetrically on the indices. Similarly, in the Fermion case, they can be written as the sum of elements of the form

$$b_{u_1\cdots u_r}\, dy^{i_1} \wedge \cdots \wedge dy^{u_r} \quad,$$

with coefficients depending symmetrically on the indices.

Of course, one can also consider $\mathcal{D}$ as mixed <u>Boson-Fermion</u>, generated by (x^i, y^u), with the following relations (in addition to the ones given above):

$$x^i y^u = y^u x^i$$

$$dx^i y^u = y^u dx^i$$

$$dx^i \wedge dy^u = -dy^u \wedge dx^i$$

In the recent physics literature, this situation is associated with the name "superspace".

<u>Remark</u>. B. Cenkl has remarked (private communication) that these formulas bear a striking resemblance to certain material in a paper by P. Rashevskii [33], which he calls the <u>Weyl algebra</u>. However, there is a difference in "gradation" involved In any case, there is certainly a strong relation to algebraic-topological ideas, particularly the basic work of H. Cartan [34]

I believe that the first explicit recognition of these formulas as a basis of a generalization of the differential form calculus was by B. Zumino [27], and, perhaps only implicitly, in the work of F. Berezin and his collaborators.

4. General Heuristic Principles about the Construction of Differential Form Algebras

In practice, one is given $\mathcal{A}$, and it is desired to construct $\mathcal{D}$. I know of no general, "natural" (i.e., "functorial") way of doing this. In fact, I would conjecture that there is one. Here are some heuristic principles.

Differential forms, i.e., elements of $\mathcal{D}$, should be linear combinations of elements of the form

$$a\, da_1 a_2 \wedge da_3 \wedge \cdots$$

Exterior multiplication should be defined in the obvious way, using "associativity":

$$(a da_1 a_2 \wedge da_3 \cdots) \wedge (b db_1 b_2 \wedge \cdots) = (a_1 da_1 a_2 \wedge da_3 \cdots b db_1 b_3 \cdots)$$

Exterior derivation should be defined as:

$$d(a da_1 a_2 \wedge da_3 \cdots) = da \wedge da_1 a_2 \wedge \cdots - a da_1 \wedge da_2 \wedge da_3 + \cdots$$

So far, $\mathcal{A}$ has had no particular commutivity properties. (That is why we write

$$a da_1 a_2 \wedge da_3 \cdots \quad ,$$

not simply
$$ada_1 \wedge da_2 \wedge \cdots \quad .)$$

If one assumes further commutativity properties (e.g., the "boson" and "fermion" situation considered in the previous section) there are implied relations among the differential forms. Here is the simplest situation.

Suppose
$$a_1 a_2 = \pm a_2 a_1 \tag{4.1}$$

Now, one of our basic postulates for d should be that:
$$d(a_1 a_2) = (da_1)a_2 + a_1 da_2 \quad . \tag{4.2}$$

Apply this to (4.1):
$$(da_1)a_2 + a_1 da_2 = \pm (da_2)a_1 + a_2 da_1)$$

If one wants the elements (a_1, a_2) to be independent, we should then have:
$$(da_1)a_2 = \pm a_2 da_1 \tag{4.3}$$

$$a_1 da_2 = \pm (da_2)a_1 \tag{4.4}$$

Hence,
$$-da_1 \wedge da_2 = da_2 \wedge da_1 \tag{4.5}$$

These are the basic commutation relations for differentials in the "superspace" situation.

Here is an algebraic situation we shall consider frequently. $\mathcal{A}$ is an associative subalgebra. It is a direct sum

$$\mathcal{A} = \mathcal{A}^+ \oplus \mathcal{A}^-$$

of two linear subspaces with the following commutation relations:

$$a^+ a = a a^+$$
$$\text{for } a^+ \in \mathcal{A}^+, \quad a \in \mathcal{A}$$

$$a_1^- a_2^- = -a_2^- a_1^-$$
$$\text{for } a_1^-, a_2^- \in \mathcal{A}^-.$$

$$\mathcal{A}^+ \cdot \mathcal{A}^+ \subset \mathcal{A}^+$$
$$\mathcal{A}^- \cdot \mathcal{A}^+ \subset \mathcal{A}^-$$
$$\mathcal{A}^- \cdot \mathcal{A}^- \subset \mathcal{A}^+$$

We shall suppose that $\mathcal{A}$ is a subalgebra of a differential form algebra $\mathcal{D}$, such that:

$$a^+(da) = (da)a^+$$
$$\text{for } a \in \mathcal{A}, \quad a^+ \in \mathcal{A}^+$$

$$a_1^- da_2^- = -da_2^- a_1^-$$
$$\text{for } a_1^-, a_2^- \in \mathcal{A}^-.$$

Hence

$$da^+ \wedge da = -da \wedge da^+, \quad da_1^- \wedge da_2^- = da_2^- \wedge da_1^-$$

5. The Cartan Differential Forms and Equations of Motion for Variational Problems in "Boson" Differential Geometry

As a basis for generalization, I will now briefly present material in previous work, describing the differential-geometry foundation of the calculus of variations in the usual framework, where all of the variables commute.

Let (x^μ) be coordinates of space-time, and (ϕ^a) be field variables. Choose the following range of indices and the summation convention:

$$0 \leq \mu,\nu \leq n \;;\quad 1 \leq a,b \leq m \;.$$

Introduce the following additional variables:

$$\phi^a_\mu$$

This will also be identified with symbols $\partial_\mu \phi^a$.

Let M be the space of variables (x, ϕ^a, ϕ^a_μ). Let

$$L(x,\phi,\partial\phi)$$

be a Lagrangian, i.e., a function on M. Introduce the following vector fields and differential forms on M:

$$\partial_\mu = \frac{\partial}{\partial x^\mu}$$

$$\partial_a = \frac{\partial}{\partial \phi^a}$$

$$\partial_a^\mu = \frac{\partial}{\partial \phi^a_\mu}$$

$$dx = dx^0 \wedge \cdots \wedge dx^3$$

$$dx_\mu = \partial_\mu \lrcorner\, dx$$

$$\theta^a = d\phi^a - \phi^a_\mu dx^\mu$$

Let

$$\omega = L dx + \theta^a \wedge \lambda_a \tag{5.1}$$

The λ_a are differential forms which are to be determined so that there are forms θ_a such that

$$d\omega = \theta^a \wedge \theta_a \tag{5.2}$$

To do this, calculate:

$$d(Ldx) = \partial_a(L)\theta^a \wedge dx + \partial_a^\mu(L)d\phi^a_\mu \wedge dx$$

$$d\theta^a = -d\phi^a_\mu \wedge dx^\mu$$

Hence,

$$d\omega = (\partial_a(L)\theta^a + \partial_a^\mu(L)d\phi^a_\mu) \wedge dx + (-d\phi^a_\mu \wedge dx^\mu) \wedge \lambda_a - \theta^a \wedge d\lambda_a$$

$$= \theta^a \wedge (\partial_a(L)dx - d\lambda_a) + d\phi^a_\mu \wedge (\partial_a^\mu(L)dx - dx^\mu \wedge \lambda_a)$$

This formula suggests that we choose λ_a so that

$$\partial_a^\mu(L)dx - dx^\mu \wedge \lambda_a = 0 \tag{5.3}$$

for (5.2) will then be satisfied.

In order to satisfy (5.3) set:

$$\lambda_a = \partial_a^\mu(L) \partial_\mu \lrcorner \, dx \tag{5.4}$$

Then,

$$dx^\mu \wedge \lambda_a = dx^\mu \partial_a^\nu(L) \partial_\nu \lrcorner \, dx$$

$$= \partial_a^\mu(L) dx \, ,$$

hence (5.3) is satisfied. With this choice, we have

$$\boxed{\omega = L dx + \partial_a^\mu(L) \theta^a \wedge (\partial_\mu \lrcorner \, dx)} \tag{5.5}$$

$$\boxed{d\omega = \theta^a \wedge \theta_a} \tag{5.6}$$

with:

$$\boxed{\theta_a = \partial_a(L) dx - d(\partial_a^\mu(L)) \wedge (\partial_\mu \lrcorner \, dx)} \tag{5.7}$$

(5.5)-(5.7) are, from our point of view, the basic formulas of the calculus of variations and Lagrangian field theory. The submanifolds of the form

$$x \to \phi(x)$$

that are solution submanifolds of the exterior differential system

$$\theta^a = \theta_a = 0$$

are the <u>extremals</u> of the variational problem. To see this, note that

$$\theta^a = 0$$

determines the conditions

$$\phi_\mu^a = \partial_\mu(\phi^a(x)) \quad,$$

while $\theta_a = 0$ determines the Euler-Lagrange equations themselves:

$$\partial_a(L) = \partial_\mu(\partial_a(L)(x,\phi(x),\partial\phi(x))) \qquad (5.8)$$

Thus, the θ^a might be called the <u>kinematic forms</u>, while the θ_a are the <u>dynamic forms</u>.

Notice that this way of developing Lagrangian field theory has the virtue of <u>involving only the algebra of differential forms</u>. Thus, it leads itself to the generalized differential form calculus developed above. To illustrate this point, we shall now develop an analogous theory for "fermion" fields.

6. Lagrangian Field Theory for Fermion Fields

Let $(x^\mu, \phi^a, \phi_\mu^a)$ now be generators of an associative algebra $\mathscr{A}$, with

$$\mathscr{A} = \mathscr{A}^+ \oplus \mathscr{A}^- \quad,$$

and the commutation relations described in Section 5. Suppose that:

$$x^\mu \in \mathcal{A}^+$$

$$\phi^a \in \mathcal{A}^-$$

$$\phi^a_\mu \in \mathcal{A}^-$$

Now, suppose that

$$L \in \mathcal{A}^+$$

is a <u>Lagrangian</u>. Let

$$\partial_\mu, \partial_a, \partial^\mu_a : \mathcal{A} \to \mathcal{A}$$

be the operators such that:

$$dL = \partial_\mu(L)dx^\mu + \partial_a(L)d\phi^a + \partial^\mu_a(L)d\phi^a_\mu \qquad (6.1)$$

Then,

$$0 = d(dL) = \partial_\nu \partial_\mu(L)dx^\nu \wedge dx^\mu + \partial_a \partial_\mu(L)d\phi^a \wedge dx^\mu$$

$$+ \partial^\nu_a \partial_\mu(L)d\phi^a_\nu \wedge dx^\mu + \partial_\mu \partial_a(L)dx^\mu \wedge d\phi^a$$

$$+ \partial_b \partial_a(L)d\phi^b \wedge d\phi^a + \partial^\mu_b \partial_a(L)d\phi^b_\mu \wedge d\phi^a$$

$$+ \partial_\nu \partial^\mu_a(L)dx^\nu \wedge d\phi^a_\mu + \partial^\nu \partial^\mu_a(L)d\phi^b_\nu \wedge d\phi^a_\mu$$

$$+ \partial_b \partial^\mu_a(L)d\phi^b \wedge d\phi^a_\mu \quad .$$

This gives the following commutative relations:

$$\partial_\nu \partial_\mu = \partial_\mu \partial_\nu$$

$$\partial_b \partial_a = -\partial_a \partial_b$$

$$\partial_b \partial_a = -\partial_a \partial_b$$

etc.

Notice that these are typical graded Lie algebra relations; in this case, the bracket operator of the graded Lie algebra is zero, i.e., it is **abelian**.

Further:

$$\partial_\mu(L) \in \mathcal{A}^+$$

$$\partial_a(L), \partial_a^\mu(L) \in \mathcal{A}^-$$

Set:

$$dx = dx^0 \wedge \cdots \wedge dx^3$$

$$\theta^a = d\phi^a - \phi_\mu^a dx^\mu$$

$$\omega = L\,dx + \partial_a^\mu(L)\theta^a \wedge (\partial_\mu \lrcorner\, dx)$$

$$\theta_a = \partial_a(L)dx - d(\partial_a^\mu(L)) \wedge (\partial_\mu \lrcorner\, dx)$$

Exactly as in the previous section, we can now demonstrate the following relation:

$$\boxed{d\omega = -\theta^a \wedge \theta_a} \qquad (6.2)$$

(The minus sign on the right hand side of (6.2) is the clue that we are dealing with "fermions".)

Again, the exterior differential system

$$\theta^a = \theta_a = 0$$

now determines the extremals of the variational problem of "extremizing"

$$\int L \, dx \quad .$$

Of course, it is not all clear what this means from the <u>traditional</u> calculus-of-variations point of view, since $\int L \, dx$ is not a real number, or even an element of a set with some sort of ordering. What we have done is to carry over the <u>formalism</u> of the calculus of variations. In essence, as Berezin and Marinov have pointed out [32], this formalizes Schwinger's point of view towards the calculus of variations.

Let us now turn to the simplest example.

7. <u>Linear Equations</u>

Keep the notation of Section 3, i.e., the ϕ^a are "fermion" field variables. Set:

$$L = \gamma^\mu_{ab}\phi^a_\mu\phi^b + \gamma_{ab}\phi^a\phi^b \qquad (7.1)$$

The $(\gamma_{ab}, \gamma_{ab})$ are real numbers. We can suppose without loss in generality that:

$$\gamma_{ab} = -\gamma_{ba} \quad .$$

Thus,

$$dL = \gamma^\mu_{ab}(d\phi^a\phi^b_\mu + \phi^a d\phi^b_\mu) + \gamma_{ab}(d\phi^a\phi^b + \phi^a d\phi^b) \quad .$$

Hence,

$$\partial_a(L) = -\gamma_{ab}\phi^b_\mu - 2\gamma_{ab}\phi^b$$

$$\partial^\mu_a(L) = \gamma^\mu_{ba}\phi^b$$

$$-\theta_a = (\gamma^\mu_{ab}\phi^b_\mu + 2\gamma_{ab}\phi^b)dx + \gamma^\mu_{ba}d\phi^b \wedge (\partial_\mu \lrcorner dx)$$

The "kinematic" equations

$$\theta^a = 0$$

mean that

$$\phi^a_\mu(x) = \partial_\mu \phi^a(x) \quad .$$

Thus the "dynamic" equations

$$\theta_a = 0$$

mean that:

$$(\gamma^\mu_{ab}\partial_\nu(\phi^b) + 2\gamma_{ab}\phi^b)dx + \gamma^\mu_{ba}(\partial_\nu(\phi^b)dx^\nu \wedge (\partial_\mu \lrcorner\, dx)$$

$$= (\gamma^\mu_{ab}\partial_\nu\phi^b + 2\gamma_{ab}\phi^b + \gamma^\mu_{ba}\partial_\mu\phi^b)dx \quad ,$$

i.e., the extremal equations are:

$$\boxed{(\gamma_{ab} + \gamma_{ba})\partial_\mu\phi^b + 2\gamma_{ab}\phi^b = 0} \qquad (7.2)$$

Remark. If we have chosen "Boson" instead of "Fermion" commutation relations for the ϕ^a, the first term on the left hand side of (7.2) would have been

$$(\gamma^\mu_{ab} - \gamma^\mu_{ba})\partial_\mu\phi^b$$

8. Lie Derivative and Contraction for Differential Form Algebras

Consider $\mathcal{D}$ for the moment as the differential form algebra of a manifold. Let $\mathcal{V}$ be the vector fields on the manifold. Then, each $V \in \mathcal{V}$ determines two operations on $\mathcal{D}$:

$$\theta \to \mathcal{L}_V(\theta)$$

$$\theta \to V \lrcorner\, \theta$$

called <u>Lie derivative</u> and <u>contraction</u>. They satisfy a series of identities (described in [4]) which are the essential ingredients in all calculations of traditional differential geometry. The most important is the following one (called E. and H. Cartan's Identity):

$$\mathscr{L}_V(\theta) = V \lrcorner d\theta + d(V \lrcorner \theta) \qquad (8.1)$$

In particular, as explained in [4], [10], [11], [12], it plays the fundamental role in the calculus of variations (for example, in the <u>first</u> and <u>second variational formula</u>).

In this section I shall attempt to extend these operations to the "fermion" differential algebras. We shall work within the following framework. $\mathscr{A}$ is an associative algebra; as a vector space it is a direct sum $\mathscr{A}^+ \oplus \mathscr{A}^-$, with

$$a^+ a^+ \subset a^+ \supset a^- a^-$$

$$a^- a^+ \subset a^-$$

$$a^+ a = a^+ \quad ; \quad a_1^- a_2^- = - a_2^- a_1^- \quad .$$

$\mathscr{A}$ will be a subalgebra of $\mathscr{D}$, a differential form algebra, with operations d and $\wedge$, such that:

$$d(a_1 a_2) = (da_1)a_2 + a_1 da_2$$

$$d(da \wedge \theta) = -da \wedge \theta$$

$$(da)a^+ = a^+(da)$$

$$(da_1^-)a_2^- = -a_2^- da_1^-$$

$$(da_1^+)a_2^- = a_2^- da_1^+$$

$\mathcal{A}$ generates $\mathcal{D}$, i.e., each element of can be written as a sum of the elements of the form

$$ada_1 \wedge \cdots \wedge da_r$$

<u>Definition</u>. Let $\alpha: \mathcal{A} \to \mathcal{A}$ be a linear map. A linear map $V: \mathcal{A} \to \mathcal{A}$ will be called an α-<u>derivation of</u> $\mathcal{A}$ if the following condition is satisfied:

$$V(a_1 a_2) = V(a_1)a_2 + \alpha(a_1)V(a_2) \qquad (8.2)$$

for $a_1, a_2 \in \mathcal{A}$.

Notice that such a V is a generalization of a "vector field" in ordinary differential geometry. This suggests that we investigate the possibility of a generalization of the "Jacobi bracket" operation. Let α_1, α_2 be linear maps: $\mathcal{A} \to \mathcal{A}$, and let $V_1, V_2: \mathcal{A} \to \mathcal{A}$ be α_1 and α_2-derivations. Then,

$$V_1 V_2(a_1 a_2) = V_1(V_2(a_1)a_2 + \alpha_2(a_1)V_2(a_2))$$

$$= V_1 V_2(a_1)a_2 + \alpha_1(V_2(a_1))V_1(a_2)$$

$$+ V_1(\alpha_2(a_1))V_2(a_2) + \alpha_1 \alpha_2(a_1)V_1 V_2(a_2)$$

Set:

$$V_3 = V_1 V_2 \pm V_2 V_1 \quad .$$

Then,

$$V_3(a_1 a_2) = V_3(a_1)a_2 \pm \alpha_1\alpha_2(a_1)V_3(a_2) + \alpha_1(V_2(a_1))V_1(a_2)$$
$$+ V_1(\alpha_2(a_1))V_2(a_2) \pm (\alpha_2(V_1(a_1))V_2(a_2)$$
$$+ V_2(\alpha_1(a_1))V_1(a_2))$$

This proves:

<u>Theorem 8.1</u>. Suppose the following conditions are satisfied:

$$\mp \alpha_2 V_1 = V_1 \alpha_2 \qquad (8.3)$$

$$\alpha_1 V_2 = \mp V_2 \alpha_1 \qquad (8.4)$$

Set:

$$\alpha_3 = \pm \alpha_1 \alpha_2 \qquad (8.5)$$

Then, $V_3 = V_1 V_2 \pm V_2 V_1$ is an α_3-derivation of $\mathcal{A}$.

<u>Remark</u>. This simple identity is the source of the "graded Lie algebra" structure on the "vector field" in "fermion differential geometry".

Let $V: \mathcal{A} \to \mathcal{A}$ be an α-derivation. Suppose that:

$$\alpha(a_1 a_2) = \alpha(a_1)\alpha(a_2) \qquad (8.6)$$
$$\text{for } a_1, a_2 \in \mathcal{A} \quad ,$$

i.e., α is an automorphism of the algebra $\mathcal{A}$. Extend V to a map

$$\theta \to \mathcal{L}_V(\theta) \quad ,$$

called <u>Lie derivation</u>, as follows.

$$\mathcal{L}_V(a_1 da_2) = V(a_1)da_2 + \alpha(a_1)dV(a_2)$$

$$\mathcal{L}_V(a_1 da_2 \wedge da_3) = V(a_1)da_2 \wedge da_3 + \alpha(a_1 \mathcal{L}_V(da_1 \wedge da_3)$$

$$= V(a_1)da_2 \wedge da_3 + \alpha(a_1)(dV(a_1) \wedge da_2$$
$$+ d\alpha(a_1 \wedge dV(a_2))$$

and so forth.

Notice that this way of defining $\mathcal{L}_V$ involves extending α to be a linear map $\mathcal{D} \to \mathcal{D}$ which is an automorphism with respect to the exterior algebra, and such that:

$$d\alpha = \alpha d \tag{8.7}$$

$$\mathcal{L}_V(\theta_1 \wedge \theta_2) = \mathcal{L}_V(\theta_1) \wedge \theta_2 + \alpha(\theta_1) \wedge \mathcal{L}_V(\theta_2) \tag{8.8}$$

for $\theta_1, \theta_2 \in \mathcal{D}$.

We can now <u>define</u> contraction by V so that the E. and H. Cartan identity is satisfied. Let us begin with $\theta = a$.

$$\mathcal{L}_V(a) = V(a) = V \lrcorner da \quad ,$$

i.e.,

$$V \lrcorner \, da = V(a) \tag{8.9}$$

Suppose

$$\theta = a_1 da_2 \, ,$$

$$\mathscr{L}_V(\theta) = V(a_1)da_2 + \alpha(a_1)dV(a_2)$$

$$\stackrel{?}{=} V \lrcorner \, (da_1 \wedge da_2) + d(V \lrcorner \, a_1 da_2) \tag{8.10}$$

Set:

$$\alpha'(ada_1 \wedge \cdots \wedge da_r) = (-1)^r \alpha(a_1) d\alpha(a_1) \wedge \cdots \wedge d\alpha(a_r) \tag{8.11}$$

$$V \lrcorner \, (da_1 \wedge da_2) = V(a_1)da_2 + \alpha'(da_1)V(a_2) \tag{8.12}$$

$$V(a_1 da_2) = \alpha'(a_1)V(a_2) \tag{8.13}$$

Thus, the right hand side of (8.11) is:

$$V(a_1)da_2 - d(\alpha(a_1))V(a_2) + d(\alpha(a_1)V(a_2))$$

$$= V(a_1)da_2 - d(\alpha(a_1))V(a_2) + d(\alpha(a_1))V(a_2) + \alpha(a_1)dV(a_2)$$

$$= V(a_1)da_2 + \alpha(a_1)dV(a_2) \, ,$$

which is precisely the left hand side of (8.11). We deduce that the question mark over the equality sign may be removed from (8.10). We can now state the definitive result:

Theorem 8.2. Let the map α be extended as an exterior algebra automorphism of $\mathscr{D}$ in two ways, labelled "α" and "α'", so that:

$$d\alpha = \alpha d$$

$$d\alpha' = -\alpha' d \quad.$$

(Precisely, α' is defined by (8.11), α by (8.11) with the term $(-1)^r$ removed.)

Then, Lie derivative and contraction are defined as linear maps $\mathscr{D} \to \mathscr{D}$ so that:

$$\mathscr{L}_V(\theta_1 \wedge \theta_2) = \mathscr{L}_V(\theta_1) \wedge \theta_2 + \alpha(\theta_1) \wedge \mathscr{L}_V(\theta_2) \qquad (8.14)$$

$$V \lrcorner (\theta_1 \wedge \theta_2) = (V \lrcorner \theta_1) \wedge \theta_2 + \alpha'(\theta_1) \wedge (V \lrcorner \theta_2) \qquad (8.15)$$

for $\theta_1, \theta_2 \in \mathscr{D}$.

The Cartan Family Identity (8.1) is then satisfied.

Proof. We have given the proof for $A = 0$ and 1. The completion of the proof is by induction on r, and is left to the reader as an exercise.

9. <u>One-Variable Calculus of Variations</u>, <u>Cauchy Characteristic</u>, <u>Hamilton-Jacobi Equation</u>, <u>Hamilton Equation</u>, <u>Poisson Bracket</u>, <u>etc</u>.

In his <u>Lecons sur les Invariants Integraux</u>, E. Cartan showed how two hundred years of work on the one-variable calculus of variations-Hamilton-Jacobi theory could be wrapped up in a very neat differential-geometric package. Now, Cartan's work was very cryptic and difficult to understand. To completely treat the classical material, it is best to combine Cartan's methods with those in Caratheodory's book [35]. My own work [1], [4], [10], [12], [22] covers this unification in some detail.

It is amazing to see how imperfectly this material has been assimilated in the scientific world. On the one hand, it underlies much of what one might call the "differential geometry of physics". Physicists still write extensively on material which is completely and elegantly subsumed under the differential geometric methods, using their own abysmal notation and ways of thinking. (What seems to have happened is that following the pattern established with group theory in the 1930's, they redo this stuff--and badly--under their own labels.) On the other hand, while there has been a modest revival of these ideas in the mathematical world, it has mainly been by the "global analysts" who have a predelection of their own of putting new labels on old wine, and losing a good deal of the content and simplicity in the process.

One would not object if the new labels were better. Unfortunately, it often amounts, as V. Arnold has well put it, to replacing "y = f(x)" by "f: X → Y". Another peculiarity of the mathematical world's approach can be examined in the article by Guido Stampacchia on "Hilbert's Twenty-Third Problem: Extensions of the Calculus of Variations" in the recent AMS Symposium on Hilbert's Problems [36]. Now, if one reads Hilbert's discussion, it is pretty clear that he had something in mind that approximated the Cartan-Caratheodory theory. He also would presumably have been pleased by the ramifications of the classical calculus of variations into contemporary science and technology--particularly Optimal Control Theory (where the Pontrjagin Maximal Principle is one beautiful and extremely useful reformulation of Hilbert's ideas) and the "Lagrangian field theory" method of the quantum field theorists. However, with the typical tunnel vision of today's overspecialized mathematicians, Stampacchia only seriously discusses his own interest--the relation between the calculus of variations and the theory of partial differential equations.

Now, it is clear that Cartan's methods only depend on the vector field-differential form calculus, and the identities relating vector fields and differential forms. Thus, it is quite simple to carry over this material to the mixed Boson-Fermion spaces (i.e., "superspaces", in the physicists' terminology), since, as we have seen, it is possible to carry over these differential form-vector field relations to these situations. In this section, I will go over this material. It is also possible to extend this material to cover the case where the basic associative algebra $\mathscr{A}$ has more general commutation relations. For example, in [24] I have shown how the Heisenberg equations of motion for quantum mechanics may be defined via this "differential form" formalism.

Also, standing in back of this material is the more general Cartan theory of exterior differential systems, which also has extensive ramifications into field theoretic physics. (For example, in the work of Estabrook and Wahlquist [37], one sees how it applies to Soliton theory.) Again, having available the generalized differential-geometric notions ("differential form", "vector field", etc.) makes these methods available for use in studying fermions, quantum systems, etc. I would imagine that the theory of "supergravity" (i.e., Einstein's gravitational theory generalized to the mixed Fermion-Boson spaces) would be especially amenable to treatment by these methods--I suspect that much of Cartan's work on relativity (in Part III of his Collected Works) can be readily carried over.

Here is one general situation where Cartan's methods apply directly. Let $(\mathscr{D}, d, \wedge, \mathscr{A})$ be a differential form algebra, with $\mathscr{A}$ the "zero-degree" forms, i.e., the subalgebra which generates (applying $\wedge$ and d) all of $\mathscr{D}$. Let $\mathscr{D}^1$ denote the "one-forms", i.e., those that can be written as

$$a_1 da_2 a_3 \quad .$$

Similarly, let $\mathcal{D}^2$, $\mathcal{D}^3$,... denote the second, third, ... degree forms.

Consider a $\omega \in \mathcal{D}^2$ such that

$$d\omega = 0 \quad .$$

Let us say that a set

$$(\theta_i, \theta^i)$$

($1 \leq i,j \leq n$, summation convention) of one-forms is **canonical** for ω if:

$$\omega = \theta_i \wedge \theta^i \quad . \tag{9.1}$$

The **Cauchy characteristics** of ω is then the exterior differential system

$$\theta_i = 0 = \theta^i \tag{9.2}$$

Example. The "spinning particle", **Fermion equations** (see Berezin-Marinov [32])

Suppose $\mathcal{A}$ is generated by elements (y^i, t), $1 \leq i,j \leq$ with:

$$y^i y^j = -y^j y^i \quad ,$$
$$t y^i = y^i t \quad ,$$

i.e., the (y^i) are "Fermion" coordinates. Let:

$$\omega = s_{ij} dy^i \wedge dy^j - dh \wedge dt \quad , \qquad (9.3)$$

where (s_{ij}) are constants, and h is an element of $\mathscr{A}$. In view of the <u>symmetry</u> of exterior multiplication for the differentials dy, we can suppose that the coefficients satisfy:

$$s_{ij} = s_{ji} \quad .$$

Suppose that:

$$dh = h_i dy^i + h_t dt \quad , \qquad (9.4)$$

i.e., the (h_i, h_t) are the "partial derivatives" of h with respect to the "coordinates". Thus,

$$\omega = s_{ij} dy^i \wedge dy^j - h_i dy^i \wedge dt$$

<u>Suppose</u> that the matrix (s_{ij}) is <u>non-degenerate</u>. (If it is not, the situation gets more complicated, but can be handled as described in [10].) Let s^{ij} be its inverse matrix. Set:

$$\theta^i = dy^i + \frac{1}{2} s^{ij} h_j dt \qquad (9.5)$$

$$\theta_i = s_{ij} \theta^j \qquad (9.6)$$

Thus,

$$\theta^i \wedge \theta_i = (dy^i + \tfrac{1}{2}s^{ij}h_j dt) \wedge (s_{ik}dy^k + \tfrac{1}{2}h_i dt)$$

$$= s_{ik}dy^i \wedge dy^k + \tfrac{1}{2}h_j dt \wedge dy^j + \tfrac{1}{2}dy^i h_i \wedge dt$$

This will equal ω provided

$$h \in \mathcal{A}^+ \tag{9.7}$$

In this case, the Cauchy characteristic curves of ω can be written as follows:

$$\frac{dy^i}{dt} = -\tfrac{1}{2} s^{ij} h_j(y(t)) \tag{9.8}$$

Notice that these are the <u>gradient curves</u> of h with respect to the (usual) <u>Riemannian</u> metric

$$s_{ij} dx^i dx^j \quad .$$

Recall the significance of these equations of electrical curcuit theory, via the work of Brayton and Moser [34]. This indicates that "Fermion" spaces have a useful role to play in Systems Theory.

<u>The Poisson bracket formalism in Fermion space</u>

Again, let $\mathcal{A}$ be the associative algebra generated by the (y^i). Let

$$\omega = s_{ij} dy^i \wedge dy^j$$

Given $a \in \mathcal{A}$, let V_a be the linear map $\mathcal{A} \to \mathcal{A}$ such that:

$$da = V_a \lrcorner \omega \quad .$$

(If $a \in \mathcal{A}^+$, choose V_a to be a derivation. If $a \in \mathcal{A}^-$, choose V_a to be an α-derivation, where $\alpha: \mathcal{A} \to \mathcal{A}$ is the isomorphism such that

$$\alpha(y^i) = -y^i \quad .)$$

Then, set

$$\{a_1, a_2\} = V_{a_1}(a_2)$$

$$\equiv \text{ the } \underline{\text{Fermion Poisson bracket}} \text{ of } a_1 \text{ and } a_2$$

The properties of this operation--which parallel the usual Poisson bracket--have been extensively discussed by Berezin [31]. In particular, it defines a <u>graded Lie algebra structure</u>.

10. <u>The Dirac Equation and Poisson Bracket of Fermion Spaces</u>

It is now recognized that Lie algebra theory plays the key role in the usual Schrödinger equation quantization procedure. (This was, of course, clearly seen by Dirac, with his prescription--Poisson brackets go into operator commutator.) One would expect that developing the parallel ideas for <u>graded Lie algebras</u> would significantly extend this "quantization" material, particularly towards a better understanding of the geometric nature of the <u>Dirac equation</u>. Berezin and Marinov [32] have already developed some of this material; my aim here is to illustrate the formalism developed with this example. Let

$$x^\mu \, ,$$

$0 \leq \mu, \nu \leq 3$ be space-time coordinates,

and let

$$\partial_\mu = \frac{\partial}{\partial x^\mu} \, .$$

Let $(g_{\mu\nu})$ be the Lorentz metric tensor. Let $g^{\mu\nu}$ be its inverse matrix. Let γ^μ be the 4×4 Dirac matrices:

$$\gamma^\mu \gamma^\nu + \gamma^\nu \gamma^\mu = \frac{1}{2} g^{\mu\nu} \tag{10.1}$$

Let

$$D = i\gamma^\mu \partial_\mu \, , \tag{10.2}$$

the <u>Dirac operator</u>. It acts on C^∞ mapping $R^4 \to C^4$.

Let $\mathscr{A}$ be the associative algebra generated by symbols

$$(x^\mu, p_\nu, y^\mu) \, ,$$

satisfying the following commutation relations:

(x^μ, p_ν) are "Boson" variables

(y^μ) are "Fermion" variables

(x,p) commute with (y) .

Aside from these relations, suppose that these generators are free. An element $\mathscr{A}$ is then of the form:

$$a = a_0(x,p) + a_\mu(x,p)y^\mu + a_{\mu\nu}(x,p)y^\mu y^\nu + \cdots$$

where the coefficients (a) are symmetric polynomials on the (x,p), and which depend skew-symmetrically on the indices. Now, set:

$$\omega = dp_\omega \wedge dx^\nu + g_{\mu\nu}dy^\mu \wedge dy^\nu \qquad (10.3)$$

Now, let $\alpha': \mathscr{D} \to \mathscr{D}$ be the algebra isomorphism such that:

$$\alpha'(y^\mu) = -y^\mu$$

$$\alpha'(x^\mu) = x^\mu$$

$$\alpha'(p_\mu) = p_\mu$$

$$d\alpha' = -\alpha'd$$

For $a \in \mathscr{A}$, let V_a be the linear map $\mathscr{A} \to \mathscr{A}$ such that

$$da = V_a \lrcorner \omega \qquad (10.4)$$

where V_a is a derivation of $\mathscr{A}$ if $a \in \alpha(a) = a$, an α-derivation of $\alpha(a) = -a$.

Now,

$$dy^\mu = V_{y^\mu} \lrcorner \omega$$

$$= g_{\nu\nu'}V_{y^\mu}(y^\nu)dy^{\nu'} + g_{\nu\nu'}\alpha'(dy^\nu)V_{y^\mu}(y^{\nu'})$$

$$= g_{\nu\nu'} V_{y^\mu}(y^\nu) dy^{\nu'} + g_{\nu\nu'} dy^\nu V_{y^\mu}(y^{\nu'})$$

Thus, we have:

$$\boxed{V_{y^\mu}(y^\nu) = \tfrac{1}{2} g^{\mu\nu}} \qquad (10.5)$$

This is the basic relation for the Dirac equation. It says that the <u>Poisson bracket</u> relations of y^μ and y^ν with respect to the two-form ω are precisely the <u>commutation relations of the Dirac matrices</u>. Thus, the Dirac differential operator D is just the result of "quantization" (following the obvious generalization of the usual "Schrödinger" quantization rules) of the element

$$p_\mu y^\nu$$

of $\mathcal{A}$.

<u>Remark</u>. This material gets more interesting if the $g_{\mu\nu}$ depend on x. This corresponds to the "Dirac equation in a Riemannian manifold" [18]. I plan to come back to this in a later publication.

Bibliography

1. R. Hermann, "Some Differential Geometric Aspects of the Lagrange Variational Problem", Illinois J. Math. 6 (1962), pp. 634-673.

2. R. Hermann, "Cartan Connections and the Equivalence Problems for Geometric Structures", Contributions to Differential Equations 3 (1964), pp. 199-248.

3. R. Hermann, "E. Cartan's Geometric Theory of Partial Differential Equations", Advances in Math. 1 (1965), pp. 265-317.

4. R. Hermann, Differential Geometry and the Calculus of Variations, Academic Press, New York, 1969.

5. R. Hermann, "A Geometric Formula for Current Algebra Commutation Relations", Phys. Rev. 177 (1969), p. 2449.

6. R. Hermann, "Quantum Field Theories with Degenerate Lagrangians", Phys. Rev. 177 (1969), p. 2453.

7. R. Hermann, Fourier Analysis on Groups and Partial Wave Analysis, W.A. Benjamin, New York, 1969.

8. R. Hermann, "Current Algebra, Sugawara Model and Differential Geometry", J. Math. Phys. 11 (1970), pp. 1825-1829.

9. R. Hermann, "Infinite Dimensional Lie Algebra and Current Algebra", Proc. 1969 Battele-Seattle Recontres on Math. Physics, Springer-Verlag, Berlin, 1970, pp. 312-337.

10. R. Hermann, Lie Algebras and Quantum Mechanics, W.A. Benjamin, New York, 1970.

11. R. Hermann, Vector Bundles in Mathematical Physics, Parts I and II, W.A. Benjamin, New York, 1970.

12. R. Hermann, Geometry, Physics and Systems, Marcel Dekker, New York, 1973.

13. R. Hermann, Physical Aspects of Lie Group Theory, University of Montreal Press, Montreal, 1974.

14. R. Hermann, Energy-Momentum Tensors, Vol. IV of Interdisciplinary Mathematics, Math Sci Press, Brookline, Mass., 1973.

15. R. Hermann, Topics in General Relativity, Vol. V of Interdisciplinary Mathematics, Math Sci Press, Brookline, Mass., 1973.

16. R. Hermann, Topics in the Mathematics of Quantum Mechanics, Vol. VI of Interdisciplinary Mathematics, Math Sci Press, Brookline, Mass., 1973.

17. R. Hermann, Geometric Structure Theory of Systems-Control Theory and Physics, Part A, Vol. IX of Interdisciplinary Mathematics, Math Sci Press, Brookline, Mass., 1975.

18. R. Hermann, Gauge Fields and Cartan-Ehresmann Connections, Part A, Vol. X of Interdisciplinary Mathematics, Math Sci Press, Brookline, Mass., 1975.

19. R. Hermann, "The Pseudopotentials of Estabrook and Wahlquist, the Geometry of Solitons, and the Theory of Connections", Phys. Rev. Lett. 36 (1976), p. 835.

20. R. Hermann, "The Deformation of Quantum into Classical Mechanics, the Poisson-Moyal Bracket, and the Classical Theory of Invariants", Harvard University Physics Dept. preprint, 1976.

21. R. Hermann, "The Inverse Scattering Technique of Soliton Theory, Lie Algebras, the Quantum Mechanical Poisson-Moyal Bracket, and the Rotating Rigid Body", Phys. Rev. Lett. 37 (1976), p. 1591.

22. R. Hermann, Geometric Structure of Systems-Control Theory and Physics, Part B, Vol. XI, Interdisciplinary Mathematics, Math Sci Press, Brookline, Mass., 1976.

23. R. Hermann, Geometry of Non-Linear Differential Equations, Bäcklund Transformations and Solitons, Part A, Vol. XII, Interdisciplinary Mathematics, Math Sci Press, Brookline, Mass., 1976.

24. R. Hermann, "Quantum and Fermion Differential Geometry à la Cartan", Harvard Physics Dept. preprint, 1977.

25. J. Wess and B. Zumino, Nucl. Phys. B78 (1974), p. 1.

26. J. Wess and B. Zumino, "Superspace Formulation of Supergravity", to appear, Physics Letters.

27. B. Zumino, in Proceedings of the Conference on Gauge Theories and Modern Field Theory, Northeastern University, 1975, eds. R. Arnowett and P. Nath (MIT Press).

28. L. Corwin, Y. Ne'eman, and S. Sternberg, Reviews of Modern Physics 47 (1975), p. 573.

29. B. Kostant, Graded Lie Algebras, preprint, MIT Math. Dept. 1976

30. F. Berezin and G.I. Kac, MAT.SB.USSR 82, 124 (1970); English translation 1970 (11), 311.

31. F. Berezin, Quantization, Alta Universetates Wrateslaviensis, no. 368.

32. F. Berezin and M. Marinov, "Particle Spin Dynamics as the Grassman Variant of Classical Mechanics", Annals of Phys. 104 (1977), 336-361.

33. P. Rashevskii, "The Real Cohomology of Homogeneous Spaces", Russian Math. Surveys, 24 (1969), 23-95.

34. H. Cartan, "Notions d'Algebre Differentielle", Colloque de Topologie, Bruxelles, 1950.

35. C. Caratheodory, Calculus of Variations and Partial Differential Equations of the First Order, Holden-Day, San Francisco, 1967.

36. F. Browder, "Mathematical Developments from the Hilbert Problems", Proc. Symp. Pure Math., Am. Math. Soc., 1974.

37. F. Estabrook and H. Wahlquist, J. Math. Phys. 16 (1), 1975.

38. R. Brayton and J. Moser, "Theory of Non-Linear Networks", Quart. Appl. Math. 22 (1964), pp. 1-33, 81-104.

VII.

DIFFERENTIAL FORMS FOR CERTAIN ASSOCIATIVE ALGEBRAS

1. INTRODUCTION

"Differential forms" are typically hybrid geometric-algebraic objects. They were developed and extensively used in the 19-th century, then fell into disrepute, partly because of the rise of Tensor Analysis (which seemed more general) and partly because they did not seem to have a "rigorous" foundation. The problem is always to make sense of "dx"--is it to be "an infinitesimal change is x?" If so, how does one make sense of this in terms of the sharpened intellectual tools introduced into mathematics with the 20-th century?

These questions did not seem to bother Elie Cartan, who (virtually alone) continued to use the 19-th century formalism (with his own touches), developing it into a tool for doing differential geometry of almost overwhelming brilliance and insight. Later on, in the 1930's and 1940's, the theory of "differential manifolds" was developed, and the classical theory of differential forms found some sort of home there. However, this modern work has involved more the dual objects--vector fields--and Cartan's methods have never been fully revived in his original form--mostly, they have been translated or dualized into the "vector field" framework.

Recent work by physicists has suggested an extension of differential geometric ideas, in order to deal as naturally with "fermions" as standard differential geometry does with "bosons". My aim here is to show that this goal can be achieved by developing Cartan's approach in certain directions he did not contemplate--particularly the assignment of objects to be called differential forms to <u>arbitrary associative algebras</u>. The traditional differential-geometric framework will reappear when this algebra is specialized to the case of the commutative algebra of C^∞, real-valued functions on a manifold. However, other possibilities may be envisaged--one gives the physicists' "fermion" differential geometry, the other gives <u>quantum mechanics</u>. The aim of this chapter is to describe the basic algebraic ideas which lead to these goodies. First, we shall--as motivation--describe the now standard description of differential forms <u>on manifolds</u>, and show how it can be reformulated to lead to the generalizations we have in mind.

2. DIFFERENTIAL FORMS ON A MANIFOLD

Let M be a manifold. F(M) denotes its C^∞, real-valued functions. Since such functions can be added and multiplied in the usual way, F(M) forms an <u>associative commulative algebra over the real numbers</u>.

A <u>vector field</u>, A, on M is an R-linear map

$$A: F(N) \to F(M)$$

which is a derivation, i.e., such that

$$A(f_1 f_2) = A(f_1)f_2 + f_1 A(f_2)$$

for $f_1, f_2 \in F(M)$

V(M) denotes the set of such vector fields. Since they can be multiplied by functions, they form an F(M)-module. The Jacobi bracket of two vector fields is again one, so the V(M) defines a Lie algebra.

The <u>differential forms</u> on M are vow (following the standard ideas) constructed from the F(M)-module V(M) using the concepts of multilinear algebra. For each integer $r \geq 1$, an r-differential form ω is an r-multilinear map

$$V(M) \times \cdots \times V(M) \to F(M)$$

which depends <u>skew-symmetrically</u> on its arguments.

There are two equations defined on differential forms-- <u>exterior multiplication</u> and <u>derivation</u>, denoted by $\wedge$ and d. $\wedge$ is an F(M)-bilinear map

$$F^r(M) \times F^s(M) \to F^{r+s}(M)$$

d is an R-<u>linear</u> map

$$F^r(M) \to F^{r+1}(M)$$

$F^0(M)$ is identified with $F(M)$ itself. Exterior multiplication is an extension of the multiplication given on $F(M)$, so that if $f \in F(M) \equiv F^0(M)$, $\omega \in F^r(M)$,

$f \wedge \omega$ will mean $f\omega$

Here are some basic properties that we will try to carry over to more general situations:

$$d = 0 \qquad (2.1)$$

Each $F^r(M)$ is generated by $F(M)$ under the operations of d and $\wedge$, i.e., each $\omega \in F^r(M)$ can be written as linear combinations of elements of the form (2.2)

$$f df_1 \wedge \cdots \wedge df_r$$

for $f, f_1, \ldots, f_r \in F(M)$

$$d(f\omega) = df \wedge \omega + f d\omega \qquad (2.3)$$

for $f \in F(M)$, $\omega \in F^r(M)$

$\wedge$ is <u>associative</u>, i.e.,

$$\omega_1 \wedge (\omega_2 \wedge \omega_3) = (\omega_1 \wedge \omega_2) \wedge \omega_3 \qquad (2.4)$$

Notice that these rules completely determine the algebraic structure of the differential forms on a manifold. In turn, using ideas originated by Cartan (and developed in "modern", i.e., post-World War II differential geometry) one can formulate all differential geometric ideas in terms of the algebra of differential forms! Thus, we have the possibility of <u>extending</u> differential-geometric ideas beyond the area within which we started (i.e., finite dimensional differentiable manifolds), provided we are given algebraic objects that possess the properties (2.1)-(2.4).

Now, this process of generalization is a standard "game" of modern mathematics, and it often has a bad name ("generalized nonsense") because it turns out to be sterile. (Speaking as a differential geometer, I might say that this is because the underlying "geometric" reality is usually neglected or distorted.) In fact, I (and presumably other people) have been aware of rich possibilities of generalization for a long time, but did not see any particular reason (at least a "geometric" reason) for its elaboration. However, the physicists have now provided us with such a motivation--the <u>study of fermions with the same</u> (or parallel) <u>geometric methods that are used to study bosons</u>. Of course, this should be no surprise to anyone familiar with the history of mathematics! Perhaps we can even be outrageous, and say that the true purpose of physics is to provide motivation for doing new mathematics!

In order to prepare the way for these generalizations, I will now develop some general ideas and motivation.

3. GENERALITIES ABOUT GRADED ALGEBRAS

First, let us recall certain standard algebraic ideas. Fix either the real or complex numbers as scalar fields. $\mathcal{A}$ is an <u>algebra</u> if it is a vector space, and is given a bilinear map

$$\mathcal{A} \times \mathcal{A} \to \mathcal{A} \ . \tag{3.1}$$

This "multiplication" is denoted in various ways. For our purposes it will be convenient to denote it by the symbol $\wedge$ used for exterior multiplication. Thus,

$$A_1 \wedge A_2$$

for $A_1, A_2 \in \mathcal{A}$

denotes the image under (A_1, A_2) under the map (3.1). The algebra is said to be <u>associative</u> if

$$A_1 \wedge (A_2 \wedge A_3) = (A_1 \wedge A_2) \wedge A_3 \qquad (3.2)$$

for $A_1, A_2, A_3 \in \mathcal{A}$

Such an algebra is said to be Z^+-<u>graded</u> if it is written (as a vector space) as a direct sum

$$\mathcal{A} = \mathcal{A}^0 \oplus \mathcal{A}^1 \oplus \mathcal{A}^2 \oplus \cdots \qquad (3.3)$$

of linear subspaces (labelled by the positive integers) such that:

$$\mathcal{A}^n \wedge \mathcal{A}^m \subset \mathcal{A}^{n+m} \qquad (3.4)$$

for each integer n, m ,

This rule tells us that $\mathcal{A}^0$ is a <u>subalgebra</u>, i.e., closed under this multiplication. Let us denote its multiplication by simple juxtaposition, i.e., the product of $a_1, a_2 \in \mathcal{A}^0$ is

$$a_1 a_2 \quad .$$

(This choice of notation is, of course, motivated by the differential form-on-a-manifold situation.)

4. DIFFERENTIAL FORM ALGEBRAS

Let $\mathcal{A} = (\mathcal{A}^n: n = 0,1,...)$ be an associative graded algebra, as described in Section 3, with the multiplication denoted by $\wedge$, reducing to juxtaposition for objects of gradation zero. Let

$$d: \mathcal{A} \to \mathcal{A}$$

be a linear map.

<u>Definition</u>. The pair $(\mathcal{A},d)$ is said to constitute a <u>differential form algebra</u> if the following conditions are satisfied:

$$dd = 0 \tag{4.1}$$

$$d(\mathcal{A}^n) \subset \mathcal{A}^{n+1} \tag{4.2}$$

$$d(a\omega) = (da)\omega + a(d\omega) \tag{4.3}$$

for $a \in \mathcal{A}^0$, $\omega \in \mathcal{A}$.

$\mathcal{A}$ is generated from $\mathcal{A}^0$ by operations of d and $\wedge$, i.e., each element of $\mathcal{A}^n$ can be written as a linear combination of forms of the type (4.4)

$$ada_1 \wedge \cdots \wedge da_n$$

with $a, a_1, \ldots, a_n \in \mathcal{A}^0$.

$\mathcal{A}$ has a unit element, "1", which belongs to $\mathcal{A}^0$. A <u>scalar</u> is a multiple (by the grand field) of 1. (4.5)

$$da = 0, \quad \text{for } a \in \mathcal{A}^0 \iff a \text{ is a scalar} \qquad (4.6)$$

We know at least one example of such a structure--the differential forms on a manifold--and also many interesting facts and concepts about this example. We shall see below that there are other interesting examples <u>motivated by physical concepts</u>. These situations make available powerful intellectual tools for building new theories!

We can now combine (4.1) and (4.3) to give the following identity:

$$d(da \wedge \omega) = -da \wedge d\omega \qquad (4.7)$$
$$\text{for } a \in \mathcal{A}^0, \quad \omega \in \mathcal{A}$$

5. COMMUTATIVITY PROPERTIES

If $\mathcal{A}$ is a differential form algebra as defined in Section 4, notice that the $\mathcal{A}^m$ for $n \geq 1$ are determined by the algebra $\mathcal{A}^0$, and the properties of d and $\wedge$. So far, we have made no particular assumptions about $\mathcal{A}^0$, beyond associativity. In the differential geometric situation (where $\mathcal{A}^0 = F(M)$), $\mathcal{A}^0$ is <u>commutative</u>. In the recent work by physicists on "fermion" spaces, $\mathcal{A}^0$ is "skew" commutative, i.e., it is generated by elements which either <u>commute</u> or <u>anitcommute</u>. We shall now investigate implications of this for the differential form algebra. Suppose $a_1, a_2 \in \mathcal{A}^0$, and

$$a_1 a_2 = \pm a_2 a_1 \quad . \tag{5.1}$$

Apply (4.3):

$$(da_1)a_2 + a_1 da_2 = \pm (da_2 a_1 + a_2 da_1) \quad .$$

Then, relation (5.1) is <u>consistent</u> with the following relations:

$$\begin{aligned} a_1 da_2 &= da_2 a_1 \\ a_2 da_1 &= a_2 da_1 \end{aligned} \tag{5.2}$$

Applying d to both sides gives now:

$$da_1 \wedge da_2 = \mp da_2 \wedge da_1 \tag{5.3}$$

This suggests the following:

<u>Definition</u>. The differential form algebra $\mathcal{A}$ is said to be <u>abelian</u> if the following conditions are satisfied

For $a_1, a_2 \in \mathcal{A}^0$,

$$\boxed{\begin{aligned} a_1 a_2 &= a_2 a_1 \\ a_1 da_2 &= da_2 a_1 \\ da_1 \wedge da_2 &= - da_2 \wedge da_1 \end{aligned}} \tag{5.4}$$

It is said to be <u>Grassmanian</u> if the following conditions are satisfied:

$\mathcal{A}^0$ is a (vector space) direct sum $\mathcal{A}^+ \oplus \mathcal{A}^-$ of linear subspaces such that:

$$\begin{aligned}
a_1^+ a_2^+ &= a_2^+ a_1^+ \in \mathcal{A}^+ \\
a_1^- a_2^+ &= a_2^+ a_1^- \in \mathcal{A}^- \\
a_1^- a_2^- &= -a_2^- a_1^- \in \mathcal{A}^+ \\
da_1^+ a_2^+ &= a_2^+ da_1^+ \\
da_1^- a_2^+ &= a_2^+ da_1^- \\
da_2^+ a_1^- &= a_1^- da_2^+ \\
a_1^- da_2^- &= -da_2^- a_1^- \\
da_1^+ \wedge da_2^+ &= -da_2^+ \wedge da_1^+ \\
da_1^- \wedge da_2^+ &= -da_2^+ \wedge da_1^- \\
da_1^- \wedge da_2^- &= da_2^- \wedge da_1^-
\end{aligned}$$

(5.5)

for $a_1^+, a_2^+ \in \mathcal{A}^+$; $a_1^-, a_2^- \in \mathcal{A}^-$.

6. THE BOSON DIFFERENTIAL FORM ALGEBRA

Here is a situation suggested by physics. Let $i, j, k, \ldots$ be indices running from one to n with summation convention. Suppose first that $\mathscr{A}$ is abelian, and generated by elements

$$x^i \tag{6.1}$$

They commute, hence

$$dx^i \wedge dx^j = -dx^j \wedge dx^i \tag{6.2}$$

$\mathscr{A}^0$ is then an abelian algebra. Multiplication by $\mathscr{A}^0$ makes each $\mathscr{A}^r$ into an $\mathscr{A}^0$-module.

Definition. $\mathscr{A}$ is said to be a <u>Boson differential form algebra</u> if the following conditions are satisfied:

> The elements (x^i) satisfy no relation except for commutativity and associativity, i.e., $\mathscr{A}^0$ is the polynomial algebra in $x^1, \ldots, x^n$. The elements
>
> $$da^{i_1} \wedge \ldots \wedge da^{i_n}$$
>
> $$1 \leq i_1 < i_2 < \cdots < i_r \leq n$$
>
> form a basis for each $\mathscr{A}^r$ as an $\mathscr{A}^0$-module.

Remark. $\mathscr{A}$ is then also isomorphic to the differential forms (in the manifold sense) on R^n whose coefficients are polynomials in the coordinates.

Thus, each $\omega \in \mathcal{A}^r$ can be written in the form:

$$\omega = a_{i_1 \cdots i_r} dx^{i_1} \wedge \cdots \wedge dx^{i_r} ,$$

with coefficients $a_{(\ )}$ which lie in $\mathcal{A}^0$, are uniquely determined by ω, and are **skew-symmetric** in the indices.

7. THE FERMION DIFFERENTIAL FORM ALGEBRA

Here is a less traditional example of a differential form algebra. Suppose $\mathcal{A}^0$ is generated by elements y^i, $1 \leq i,j \leq n$, such that:

$$y^i y^j = - y^j y^i \qquad (7.1)$$

$$\text{for } 1 \leq i,j \leq n .$$

Thus, every $a \in \mathcal{A}^0$ can be written as the sum of elements of the form

$$k_{i_1 \cdots i_r} y^{i_1} \cdots y^{i_r} \qquad (7.2)$$

with coefficients k_1, from the scalar field which depend **skew-symmetrically** on the indices. We shall suppose that the elements (y^i) are free from relations in the sense that the elements

$$y^{i_1} \cdots y^{i_r} \qquad (7.3)$$

$$1 \leq i_1 < \cdots < i_r, \leq n$$

are linearly independent.

Exercise. Show that this is so if and only if $\mathcal{A}^0$ --as a vector space over the scalar field--has dimension 2^n.

$\mathcal{A}^0$, defined in this way, will be called the <u>Fermion algebra</u>. Let us <u>suppose</u> that $\mathcal{A}^0$ is part of a differential form algebra $(\mathcal{A}^n) = \mathcal{A}$, of Grassmanian type. The commutation relation (7.1) then requires the following commutation relations for the differentials

$$y^i dy^j = -dy^j y^i \qquad (7.4)$$

$$dy^i \wedge dy^j = dy^j \wedge dy^i \ . \qquad (7.5)$$

Thus, each $\omega \in \mathcal{A}^n$ can be written as sum of elements of the form:

$$a_{i_1 \cdots i_n} dy^{i_1} \wedge \cdots \wedge dy^{i_n} \ , \qquad (7.6)$$

where $a_1 \ldots \in \mathcal{A}^0$, and they depend <u>symmetrically</u> on the indices. We shall say that $\mathcal{A}$ forms the <u>Fermion differential form algebra</u> if the elements

$$dy^{i_1} \wedge \cdots \wedge dy^{i_n}$$

$$1 \leq i_1 \leq i_2 \leq \cdots \leq i_r \leq n$$

are linearly independent relative to $\mathcal{A}^0$, i.e., if no linear combination of elements of form (7.6) can vanish without the coefficients vanishing.

Supposing that the Fermion differential form algebra exists, we have seen what it looks like. Conversely:

Exercise. Show that it exists.

Notice the interesting algebraic way the Boson differential form algebra differs from the Fermion one. The Bosons are essentially "tensors" of the form

$$\tau = \left(k_{i_1\cdots i_r, j_1\cdots j_n}\right)$$

which depend symmetrically on the i-indices, skew-symmetrically on the j-indices. To see this, identify τ with

$$k_{i_1\cdots i_r, j_1\cdots j_n} x^{i_1}\cdots x^{i_r} dx^{i_1}\wedge\cdots\wedge dx^{i_r}$$

The "Fermions" are obtained by requiring the reverse, i.e., skew-symmetry in the i's, symmetry in the j's. It will be interesting to correlate the physical properties with these mathematical ones.

8. A CHANGE IN NOTATION WITH $\mathcal{A}$ REPLACING $\mathcal{A}^0$, $\mathcal{D}^r(\mathcal{A})$ REPLACING $\mathcal{A}^r$

The notation used up to now was designed to explain certain foundational points as efficiently as possible. It is not the most convenient notation to use for further developments. Accordingly, I will suggest and use an alternative one.

We have seen that the whole differential form algebra is often determined by algebra $\mathscr{A}^0$ and its algebraic properties. Accordingly, let us call it "$\mathscr{A}$", "$\mathscr{B}$", "$\mathscr{F}$", etc. For the moment, stick with $\mathscr{A}$. Denote the r-th degree differential forms by $\mathscr{D}^r(\mathscr{A})$. Thus,

$$\mathscr{D}^0(\mathscr{A}) = \mathscr{A} \ .$$

We shall say that $\mathscr{D}(\mathscr{A}) = \{\mathscr{D}^r(\mathscr{A})\}$ is the differential form algebra based on the algebra $\mathscr{A}$, assuming that some specific way (analogous to what we have just done for the Boson and Fermion case) has been chosen to make it unique.

9. THE Z_2-GRADED STRUCTURE FOR THE BOSON DIFFERENTIAL FORM ALGEBRA

Let $\mathscr{A}$ be a commutative associative algebra, and let $\mathscr{D}^r(\mathscr{A})$, $r = 0,1,...$, be the differential form algebra based on it. $\mathscr{D}(\mathscr{A})$ denotes the direct sum of all the $\mathscr{D}^r(\mathscr{A})$. We shall define linear subspace $\mathscr{D}^+(\mathscr{A})$, $\mathscr{D}^-(\mathscr{A})$ of $\mathscr{D}(\mathscr{A})$ such that

$$\mathscr{D}(\mathscr{A}) = \mathscr{D}^+(\mathscr{A}) \oplus \mathscr{D}^-(\mathscr{A}) \ ,$$

i.e., we shall define a Z_2-graded structure. It will have certain interesting properties relative to exterior derivative and multiplication.

We shall, of course, suppose that

$$a_1 da_2 = da_2 a_1 \ , \tag{9.1}$$

$$da_1 \wedge da_2 = -da_2 \wedge da_1 \qquad (9.2)$$

for $a_1, a_2 \in \mathcal{A}$.

Exercise. Show that the commutation relations (9.1)-(9.2) imply the following relations:

$$\omega_1 \wedge \omega_2 = (-1)^{nm} \omega_2 \wedge \omega_1 \qquad (9.3)$$

for $\omega_1 \in \mathcal{D}^n(\mathcal{A})$, $\omega_2 \in \mathcal{D}^m(\mathcal{A})$

$$d(\omega_1 \wedge \omega_2) = d\omega_1 \wedge \omega_2 + (-1)^n \omega_1 \wedge d\omega_2 \qquad (9.4)$$

((9.3)-(9.4) are the familiar rules for the differential forms that occur in differential geometry.)

These rules suggest that we make the following definitions:

$$\mathcal{D}^+(\mathcal{A}) = \mathcal{D}^0(\mathcal{A}) \oplus \mathcal{D}^2(\mathcal{A}) + \cdots \qquad (9.5)$$

$$\mathcal{D}^-(\mathcal{A}) = \mathcal{D}^1(\mathcal{A}) \oplus \mathcal{D}^3(\mathcal{A}) + \cdots \qquad (9.6)$$

Thus we have the following rules:

$$d(\mathcal{D}^+(\mathcal{A})) \subset \mathcal{D}^-(\mathcal{A}) \qquad (9.7)$$

$$d(\mathcal{D}^-(\mathcal{A})) \subset \mathcal{D}^+(\mathcal{A}) \qquad (9.8)$$

$$\mathcal{D}^+(\mathcal{A}) \wedge \mathcal{D}^+(\mathcal{A}) \subset \mathcal{D}^+(\mathcal{A}) \qquad (9.9)$$

$$\mathcal{D}^-(\mathcal{A}) \wedge \mathcal{D}^-(\mathcal{A}) \subset \mathcal{D}^+(\mathcal{A}) \qquad (9.10)$$

$$\mathcal{D}^-(\mathcal{A}) \wedge \mathcal{D}^+(\mathcal{A}) \subset \mathcal{D}^-(\mathcal{A}) \tag{9.11}$$

$$\omega_1^+ \wedge \omega_2^\pm = \omega_2^\pm \wedge \omega_1^+ \tag{9.12}$$

$$\omega_1^- \wedge \omega_2^- = -\omega_2^- \wedge \omega_1^- \tag{9.13}$$

Here is another way of looking at these rules. Define a linear map

$$\alpha: \mathcal{D}(\mathcal{A}) \to \mathcal{D}(\mathcal{A})$$

as follows:

$$\alpha(\omega^\pm) = \pm\omega \tag{9.14}$$
$$\text{for } \omega^\pm \in \mathcal{D}^\pm(\mathcal{A})$$

Then,

$$\alpha(\omega_1 \wedge \omega_2) = \alpha(\omega_1) \wedge \alpha(\omega_2) \tag{9.15}$$

$$\alpha(d\omega) = -d\alpha(\omega) \tag{9.16}$$

$$\alpha^2 = 1 \tag{9.17}$$

Then the essential algebraic core of what we have done is the construction of the triple

$$(\mathcal{D}(\mathcal{A}), d, \alpha)$$

α with these properties will be called a <u>parity operator</u> for the differential form algebra $(\mathcal{D}(\mathcal{A}), d)$.

10. THE COMMUTATION RELATIONS AND PARITY OPERATOR FOR GRASSMANIAN DIFFERENTIAL FORM ALGEBRAS

Let $\mathcal{A}$ be an associative algebra, with multiplication denoted by simple juxtaposition. Let

$$\alpha: \mathcal{A} \to \mathcal{A}$$

be an algebra automorphism such that

$$\alpha^2 = 1 \quad .$$

Set:

$$\mathcal{A}^+ = \{a \in \mathcal{A} : \alpha(a) = a\}$$

$$\mathcal{A}^- = \{a \in \mathcal{A} : \alpha(a) = -a\} \quad .$$

Then, it follows from an assumption that α be an automorphism that

$$\mathcal{A}^+ \mathcal{A}^+ \subset \mathcal{A}^+ \tag{10.1}$$

$$\mathcal{A}^- \mathcal{A}^+ \subset \mathcal{A}^- \tag{10.2}$$

$$\mathcal{A}^+ \mathcal{A}^- \subset \mathcal{A}^- \tag{10.3}$$

$$\mathcal{A}^- \mathcal{A}^- \subset \mathcal{A}^+ \tag{10.4}$$

<u>Definition</u>. We shall say that (a, α) is a <u>Grassman algebra</u> if the following commutation relations are satisfied:

$$\boxed{\begin{aligned} a_1^+ a_2^+ &= a_2^+ a_1^+ \\ a_1^- a_2^- &= - a_2^- a_1^- \\ a_1^+ a_2^- &= a_2^- a_1^+ \end{aligned}}$$

(10.5)

for $a_1^+, a_2^+ \in \mathcal{A}^+$; $a_1^- a_2^- \in \mathcal{A}^-$

Let such a Grassman algebra be given. Let us also suppose that there is a differential form algebra

$$\mathcal{D}(\mathcal{A}) = \{\mathcal{D}^n(\mathcal{A})\} ,$$

with $\mathcal{D}^0(\mathcal{A}) = \mathcal{A}$. Recall that it is said to be a <u>Grassman differential form algebra</u> if the following commutation relations are satisfied:

$$\boxed{\begin{aligned} a_1 da_2^+ &= da_2^+ a_1 \\ da_1 \wedge da_2^+ &= - da_2^+ \wedge da_1 \\ a_1^- da_2^- &= - da_2^- a_1^- \\ da_1^- \wedge da_2^- &= da_2^- \wedge da_1^- \end{aligned}}$$

(10.6)

for $a_1^+, a_2^+ \in \mathcal{A}^+$; $a_1^-, a_2^- \in \mathcal{A}^-$.

Definition. The parity operator α is said to <u>extend to</u> $\mathcal{D}(\mathcal{A})$ if there is a linear map: $\mathcal{D}(\mathcal{A}) \to \mathcal{D}(\mathcal{A})$, which we again denote by α, satisfying the following conditions:

Restricted to $\mathcal{A} \equiv \mathcal{D}^0(\mathcal{A})$, it is just the parity operator α with which we began.

$$\alpha^2 = 1 \qquad (10.7)$$

$$d\alpha = -\alpha d \qquad (10.8)$$

$$\alpha(\omega_1 \wedge \omega_2) = \alpha(\omega_1) \wedge \alpha(\omega_2) \qquad (10.9)$$

for $\omega_1, \omega_2 \in \mathcal{D}(\mathcal{A})$,

i.e., α is an <u>algebraic-automorphism</u> relative to the exterior product.

Let us now suppose that $(\mathcal{D}(\mathcal{A}), \alpha, d)$ satisfies these conditions. Let:

$$\omega = a^{\pm} da_1^{+} \wedge \cdots \wedge da_n^{+} \wedge da_1^{-} \wedge \cdots \wedge da_m^{-} \qquad (10.10)$$

Then,

$$\alpha(\omega) = \pm (-1)^n (-1)^{2m} \omega = \pm (-1)^n \omega \qquad (10.11)$$

Set:

$$\mathcal{D}^{+}(\mathcal{A}) = \{\omega \in \mathcal{D}(\mathcal{A}): \alpha(\omega) = \omega\}$$

$$\mathcal{D}^{-}(\mathcal{A}) = \{\omega \in \mathcal{D}(\mathcal{A}): \alpha(\omega) = -\omega\} \quad .$$

Thus, formula (10.9) determines which monomials of the form (10.8) belong to $\mathcal{D}^{+}$, which to $\mathcal{D}^{-}$. It also shows that α can be extended from $\mathcal{A}$ to $\mathcal{D}(\mathcal{A})$, as we have assumed. Here are the commutation rules for $\mathcal{D}(\mathcal{A})$:

Theorem 10.1.

$$\boxed{\begin{aligned} \omega_1^+ \wedge \omega_2^\pm &= \omega_2^\pm \wedge \omega^+ \\ \omega_1^- \wedge \omega_2^- &= -\omega_2^- \wedge \omega_1^- \end{aligned}} \qquad (10.12)$$

First note that (10.12) is true if ω_1, ω_2 are of degree zero, i.e., elements of $\mathcal{A}$. Let us now proceed by induction on the degree of $\omega_1^\pm$. Suppose it is an exterior product of forms of lower degree, e.g.,

$$\omega_1^\pm = \theta_1^\pm \wedge \theta_2^+ \quad .$$

Then

$$\begin{aligned}
\omega_1^\pm \wedge \omega_2^+ &= (\theta_1^\pm \wedge \theta_2^+) \wedge \omega_2^+ \\
&= \theta_1^\pm \wedge \omega_2^+ \wedge \theta_2^+ \\
&= \omega_2^+ \wedge \theta_2^\pm \wedge \theta_2^+ \\
&= \omega_2^+ \wedge \omega_1^\pm \\
\omega_1^- \wedge \omega_2^- &= (\theta_1^- \wedge \theta_2^+) \wedge \omega_2^- \\
&= \theta_1^- \wedge \omega_2^- \wedge \theta_2^+ \\
&= -\omega_2^- \wedge \omega_1^- \quad .
\end{aligned}$$

This checks with (10.12). It is left to the reader to complete the checking.

Remark. This result is trivial, but very "elegant". It suggests that the proper "category" to be carrying on all of this is that of "differential algebras equipped with a parity operator", i.e., an associate algebra $\mathscr{A}$, with a linear map d: $\mathscr{A} \to \mathscr{A}$ such that $d^2 = 0$, and a linear map $\alpha: \mathscr{A} \to \mathscr{A}$ such that $d\alpha = -\alpha d$, $\alpha^2 = 1$.

VIII.

THE POISSON-MOYAL BRACKET AND QUANTUM SYMPLECTIC GEOMETRY

1. INTRODUCTION

Let X be the space R^{2n} interpreted as the state space ($\equiv$ cotangent bundle of R^n) of a classical mechanical system with n degrees of freedom. Let (x^i), $1 \leq i,j \leq n$ be the functions on X which represent the position coordinates, and let (p_i) denote the functions which represent the momenta. Then, (x^i, p_i) define a coordinate system for X.

Let $\mathscr{F}(X)$ denote the real-valued, C^∞ functions on X. They may be considered as the <u>observables</u> of the classical mechanical system. They admit a Lie algebra structure--called <u>Poisson bracket</u>--defined as:

$$f_1, f_2 = \frac{\partial f_1}{\partial p_i} \frac{\partial f_2}{\partial x^i} - \frac{\partial f_2}{\partial p_i} \frac{\partial f_1}{\partial x^i} \quad .$$

It is a very familiar story how this is related to the <u>symplectic structure</u>

$$\omega = dp_i \wedge dx^i \qquad (1.1)$$

on X.

Now, consider "quantum mechanics". It depends on a parameter $\hbar$ that is not present in classical mechanics, <u>Planck's constant</u>. One way of looking at it is that it defines a one-parameter family

$$\{\ ,\ \}_\hbar$$

of Lie algebra structure on $\mathscr{F}(X)$, reducing to $\{\ ,\ \}$ as $\hbar \to 0$. This "new" bracket has the following form

$$\{f_1, f_2\}_\hbar = \sum_{j=0}^{\infty} \frac{\hbar^{2j}}{(2j+1)!} D_j(f_1, f_2)$$

when $D_j: \mathscr{F}(X) \times \mathscr{F}(X) \to \mathscr{F}(X)$ are bilinear differential operators called (in the 19-th century) the <u>transvections</u>. ("Uberschiebung", in German.)

In this chapter we shall first look into the definitions of these operations in more detail and second, look for their "geometric" explanation. I shall look along the following lines: Using the Heisenberg commutation relations, make $\mathscr{F}(X)$ into a <u>non-commutative</u> associative algebra. Define "differential forms" using this algebraic structure, as described in previous chapters. Thus, (1.1) might be interpreted as defining a "symplectic structure" in terms of the <u>non-commutative differential geometry</u>. This observation suggests a "program", which might lead to significant new insights into quantum mechanics (and differential geometry!).

Another interesting feature is the relation to 19-<u>th</u> <u>century invariant theory</u>. In fact, these "transvection" operators played a prominent role in that discipline. For $n = 1$, they are invariant under $SL(2,R)$ ($\equiv Sp(1,R)$), and the 19-th century work concentrated on this group. Thus, we see again the broad ramifications of quantum mechanics!

2. THE TRANSVECTION OPERATIONS

Let X continue as R^{2n}, with "canonical" coordinates denoted as (x^i, p_i), $1 \leq i, j \leq n$. Let $\mathscr{F}$ ($\equiv \mathscr{F}(X)$) denote the C^∞, real-valued functions on $\mathscr{F}$. Consider $\mathscr{F}$ as a <u>commutative</u>, <u>associative algebra</u> over the real numbers. Let

$$\partial_i \equiv \frac{\partial}{\partial x^i} \quad ; \quad \partial^i \equiv \frac{\partial}{\partial p_i}$$

be the R-linear maps: $\mathscr{F} \to \mathscr{F}$ defined by the partial derivatives with respect to the coordinates. Set:

$$\mathscr{F}^2 = \mathscr{F} \otimes \mathscr{F} \equiv \text{tensor product (over the field of \underline{real} scalars) of two copies of the \underline{vector space}} \; \mathscr{F}.$$

Let $\tau: \mathscr{F}^2 \to \mathscr{F}^2$ be the R-linear map defined by the following formula:

$$\tau(f \otimes g) = \partial^i(f) \otimes \partial_i(g) - \partial_i(f) \otimes \partial^i(g) \qquad (2.1)$$

for $f, g \in \mathscr{F}$.

In other words, in terms of the notion of "tensor product of linear maps" (IM, Vol. 2),

$$\tau = \partial^i \otimes \partial_i - \partial_i \otimes \partial^i . \qquad (2.2)$$

<u>Remark</u>. The summation convention is in force. It is also necessary to keep in mind the possible ambiguity in the scalars chosen for tensor product. Formula (2.1) only makes <u>sense</u> for tensor product over the <u>reals</u> (recall that "tensor analysis" involves "tensor products of $\mathscr{F}$-modules").

Definition. The R-linear map τ is called the __symbol__ of the transvection operation.

Let
$$M: \mathscr{F}^2 \equiv \mathscr{F} \otimes \mathscr{F} \to \mathscr{F}$$
be the R-linear map defined by algebra multiplication, i.e.,
$$M(f_1 \otimes f_2) = f_1 f_2 \quad .$$

Definition. For each integer j, the j-th transvection operation
$$T_j: \mathscr{F} \otimes \mathscr{F} \to \mathscr{F}$$
is defined by the following formula
$$T_j = M\tau^j \tag{2.3}$$
($\tau^j \equiv j$-th power of the linear map τ.)

3. SOME TRADITIONAL SPECIAL CASES OF THE TRANSVECTION OPERATION

First, take
$$\boxed{j = 1} \quad .$$

$$T_1(f \otimes g) = \partial^i(f)\partial_i(g) - \partial_i(f)\partial^i(g)$$

$$= \frac{\partial f}{\partial p_j} \frac{\partial g}{\partial x^i} - \frac{\partial f}{\partial x^i} \frac{\partial g}{\partial p_i}$$

$$= \frac{\partial f}{\partial p_i}\frac{\partial g}{\partial x^i} - \frac{\partial g}{\partial p_i}\frac{\partial f}{\partial x^i} \qquad (3.1)$$

$$\equiv \boxed{\text{Poisson bracket}}$$

$$\boxed{j = 1, \quad n = 1}$$

$$T_1(f \otimes g) = \frac{\partial f}{\partial p}\frac{\partial g}{\partial x} - \frac{\partial f}{\partial x}\frac{\partial g}{\partial p} \qquad (3.2)$$

This is the Jacobian of the map

$$R^2 \to R^2$$

defined as $(x,p) \to (f(x,p), g(x,p))$. In classical invariant theory (see Elliot [1] or Salmon [2]). The right hand side of (3.2) is called the <u>Jacobian</u> of f_1, f_2.

$$\boxed{j = 2, \quad n = 2. \quad \text{The Hessian}}$$

When computing with one or two variable calculus, it is most convenient to use the notation of nonlinear wave theory (Whitham [3]). Partial derivatives (e.g., with respect to x and p) are denoted by subscripts, i.e.,

$$\frac{\partial f}{\partial x} = f_x, \qquad \frac{\partial f}{\partial p} = f_p,$$

and so forth.

Thus, for $f, g \in \mathcal{F}$,

$$\tau(f \otimes g) = f_p \otimes g_x - f_x \otimes g_p$$

$$\begin{aligned}\tau^2(f \otimes g) &= \tau(f_p \otimes g_x) - \tau(f_x \otimes g_p) \\ &= f_{pp} \otimes g_{xx} - f_{px} \otimes g_{xp} - f_{xp} \otimes g_{px} \\ &\quad + f_{xx} \otimes g_{pp} \\ &= f_{pp} \otimes g_{xx} - 2f_{px} \otimes g_{xp} + f_{xx} \otimes g_{pp}\end{aligned}$$

Hence,

$$T_2(f,g) = f_{pp}g_{xx} - 2f_{px}g_{xp} + f_{xx}g_{pp} \tag{3.3}$$

The right hand side of (3.2) is classically called the $\boxed{\text{Hessian}}$ of f and g. It plays a prominent role in classical invariant theory.

For example, for

$$\begin{aligned}f = g &= \text{a quadratic form in } x \text{ and } p \\ &= a_{11}x^2 + a_{12}x_p + a_{22}p^2 \\ T_2(f,f) &= 4a_{22}a_{11} + 4a_{11}a_{22} - 2a_{12}^2 \\ &= 8a_{11}a_{22} - 2a_{12}^2\end{aligned} \tag{3.4}$$

Now,

$$\begin{vmatrix} 2a_{11} & a_{12} \\ a_{12} & 2a_{22} \end{vmatrix} = 4a_{11}a_{22} - a_{12}^2$$

$$\equiv \text{"discriminant" of the quadratic form}$$

Thus we see that the second transvection operator picks out the key algebraic invariant. Another way of looking at this is to set $p = 1$ (i.e., "pass to inhomogeneous coordinates"), and consider the quadratic equation

$$a_{11}x^2 + a_{12}x + a_{22} .$$

Its roots are

$$\frac{-a_{12} \pm \sqrt{a_{12}^2 - 4a_{11}a_{22}}}{2a_{11}}$$

Again, we see appear the discriminant. This is a very simple and trivial illustration of a key 19-th century invariant theory trick--construct "invariants" by looking at the "geometric" invariants of the objects determined by the algebra--in this case the nature of the locus

$$f(x,p) = 0 .$$

4. COVARIANCE PROPERTIES OF THE TRANSVECTION

In order to carry out certain computations conveniently, we now change notation. Let $X = R^n$. (Here R^n replaces R^{2n}.) (x^i), $1 \leq i,j \leq n$, are Cartesian coordinates for R^n.

$$\partial_i = \frac{\partial}{\partial x^i} .$$

Let

$$\tau: \mathscr{F} \to \mathscr{F}$$

be defined as follows:

$$\tau(f_1 \otimes f_2) = \tau^{ij} \partial_i(f_1) \otimes \partial_j(f_2) \tag{4.1}$$

for $f_1, f_2 \in \mathscr{F}$,

i.e.,

$$\tau = \tau^{ij} \partial_i \otimes \partial_j \tag{4.2}$$

(τ^{ij}) is a constant matrix.

Let $\phi: \mathscr{F} \to \mathscr{F}$ be an automorphism. (It is the algebraic analogue of a diffeomorphism $R^n \to R^n$.) Suppose

$$\phi \partial_i \phi^{-1} = J_i^j \partial_j \tag{4.3}$$

(Thus, (J_i^j) is the Jacobian matrix of the map $x \to \phi(x)$ of $R^n \to R^n$.)

We want to compute explicitly the action of ϕ on τ. Extend ϕ to act on $\mathscr{F} \otimes \mathscr{F}$ as follows

$$\phi(f_1 \otimes f_2) = \phi(f_1) \otimes \phi(f_2) ,$$

i.e., let ϕ act as "tensor product" $\phi_1 \otimes \phi_2$. Thus,

$$(\phi \tau \phi^{-1})(f_1 \otimes f_2) = \tau^{ij} \phi \partial_i \phi^{-1} f_1 \otimes \phi \partial_j \phi^{-1} f_2$$

$$= \tau^{ij} J_i^k \partial_k f_1 \otimes J_j^\ell \partial_\ell f_2 \tag{4.4}$$

Remark. Here is a subtle point in our notation. We <u>cannot</u> "pull the J's through the tensor product symbol" $\otimes$ and write the right hand side of (4.4) as

$$\tau^{ij}J_i^k J_k^\ell \partial_k(f_1) \otimes \partial_\ell(f_2)$$

if the J's are non-constant. For the tensor product $\otimes$ refers to the tensor product as a real vector space.

However, we can follow $\mathscr{F}$ with the "multiplication" operator

$$M: \mathscr{F} \otimes \mathscr{F} \to \mathscr{F} \quad,$$

obtain

$$T(f_1 \otimes f_2) = M\tau(f_1 \otimes f_2) \quad.$$

Then,

$$(\phi T \phi^{-1})(f_1 \otimes f_2) = \tau^{ij}J_i^k J_j^\ell \partial_k(f_1)\partial_\ell(f_2) \quad.$$

Hence, we have proved:

Theorem 4.1. $\phi T \phi^{-1} = T$ if and only if the Jacobian matrix J_i^j satisfies the following condition:

$$\tau^{ij}J_i^k J_j^\ell = \tau^{k\ell} \tag{4.5}$$

This allows J_i^j to be non-constant. In fact, (4.5) says that ϕ, as a map $R^n \to R^n$, acts as an isomorphism of the tensor field

$$\tau^{ij}\partial_i \otimes_{\mathscr{F}} \partial_j$$

on R^n. (Here $\otimes_{\mathscr{F}}$ denotes tensor field in the sense of $\mathscr{F}$-module, i.e., a "tensor field on the manifold R^n in the classical sense. In this formula, ∂_i is most conveniently defined as a vector field on R^n.)

Remark. If $\tau^{ij} = -\tau^{ji}$, this is the geometric structure I have called a cosymplectic structure. (See IM, Vol. 10 and Vol. 15.)

Let us look for the condition that ϕ commute with T_2. Now,

$$T_2 = M\tau^2 \ .$$

τ is of the form

$$\tau^{ij} \partial_i \otimes \partial_j$$

where τ^{ij} is a constant matrix, ∂_i are first order differential operators: $\mathscr{F} \to \mathscr{F}$. Then,

$$\tau^2(f_1 \otimes f_2) = (\tau^{ij}\partial_i \otimes \partial_j)(\tau^{k\ell}\partial_k(f_1) \otimes \partial_\ell(f_2))$$

$$= \tau^{ij}\tau^{k\ell}\partial_i\partial_k(f_1) \otimes \partial_j\partial_\ell(f_2)$$

i.e.,

$$\tau^2 = \tau^{ij}\tau^{k\ell}\partial_i\partial_k \otimes \partial_j\partial_\ell$$

Hence,

$$\phi\tau^2\phi^{-1} = \tau^{ij}\tau^{k\ell}\phi(\partial_i\partial_k)\phi^{-1} \otimes \phi(\partial_j\partial_\ell)\phi^{-1}$$

$$\phi T_2 \phi^{-1}(f_1,f_2) = \tau^{ij}\tau^{k\ell}\phi(\partial_i\partial_k)\phi^{-1}(f_1)\phi(\partial_j\partial_\ell)\phi^{-1}(f_2)$$

Suppose that

$$\phi T_2 \phi^{-1} = T_2 \tag{4.6}$$

Now,

$$\phi(\partial_i\partial_k)\phi^{-1} = J_{ik}^{i_1 k_1} \partial_{i_1}\partial_{k_1} + J_{ik}^{i_1} \partial_{i_1} \ .$$

We see that (4.6) imposes conditions on the functions $J_{ik}^{i_1}$ (which are essentially just the derivatives of the J_i^j.) Thus, the only <u>automatic</u> condition which guarantees that ϕ commute with T_2 is that ϕ commute with T_1, and that J_i^j be <u>constant</u>. The reader is invited to work out the conditions that ϕ commute with $T_2, T_3, \ldots$ in more detail.

5. THE ASSOCIATIVE MOYAL PRODUCT AND THE POISSON-MOYAL LIE ALGEBRA

Let $\mathscr{F}$ be the algebraic C^∞ functions on R^n, and let $\partial_i : \mathscr{F} \to \mathscr{F}$, $1 \leq i, j \leq n$, be the coordinate differential operators. Let τ^{ij} be a <u>constant</u> matrix. Let

$$\tau : \mathscr{F} \otimes \mathscr{F} \to \mathscr{F} \otimes \mathscr{F}$$

be the following linear map

$$\tau(f_1 \otimes f_2) = \tau^{ij} \partial_i(f_1) \otimes \partial_j(f_2) \qquad (5.1)$$

Let $M : \mathscr{F} \otimes \mathscr{F} \to \mathscr{F}$ be the multiplication map. Set

$$f_1 * f_2 = \sum_{j=0}^{\infty} M \frac{\tau^j}{j!} (f_1 \otimes f_2) \quad . \qquad (5.2)$$

(Work with (5.2) as a <u>formal</u> power series.) $f_1 * f_2$ is called the <u>Moyal Product</u>.

Let us look at the conditions that it be associative, i.e., that

$$f_1 * (f_2 * f_3) = (f_1 * f_2) * f_3 \qquad (5.3)$$

(Since (5.2) does not necessarily converge, (5.3) must be

interpreted in an appropriate way, i.e., as a relation among formal power series.) A horrendous mess results if (5.2) is substituted directly into (5.3). Let us use Fourier analysis to simplify. If $p = (p_i) \in R^n$, set

$$p \cdot x = p_i x^i$$

$$e_p(x) = e^{p \cdot x}$$

$$\tau(e_{p_1} \otimes e_{p_2}) = \tau^{ij} \partial_i(e_{p_1}) \otimes \partial_j(e_{p_2})$$

$$= \tau(p_1, p_2) e_{p_1} \otimes e_{p_2} \quad ,$$

where

$$\tau(p_1, p_2) = \tau^{ij}(p_1)_i (p_2)_j \tag{5.4}$$

Hence,

$$e_{p_1} * e_{p_2} = \exp(\tau(p_1, p_2)) e_{(p_1+p_2)} \tag{5.5}$$

(Notice that these are essentially the "Heisenberg commutation relations" in the "Weyl" form!) Then,

$$e_{p_1} * \left(e_{p_2} * e_{p_3}\right) = \exp(\tau(p_2, p_3) + \tau(p_1, p_2+p_3)) e_{p_1+p_2+p_3}$$

$$\left(e_{p_1} * e_{p_2}\right) * e_{p_3} = \exp(\tau(p_1, p_2) + \tau(p_1+p_2, p_3)) e_{p_1+p_2+p_3}$$

Hence, the condition that the associative law (5.3) be satisfied by the Moyal product is equality of the scalar coefficients, i.e.,

$$\tau(p_2,p_3) + \tau(p_1,p_2) + \tau(p_1,p_3) = \tau(p_1,p_2) + \tau(p_1,p_3) + \tau(p_2,p_3)$$

which is, of course, satisfied identically. Let us sum up as follows.

<u>Theorem 5.1</u>. If τ is a bilinear map $R^n \times R^n \to R$, the formula (5.2) defines an associative algebra structure on $\mathscr{F}$. It is called the <u>Moyal product</u>.

<u>Remark</u>. Notice that τ does not have to be non-singular. Thus, by not tying it down to be <u>the</u> form responsible for the traditional Heisenberg commutation relations (i.e., $R^n = R^{2n} \equiv$ cotangent bundle of R^m) we have received a bonus--a technique that works equally well for mechanical systems "with constraints".

Given any associative algebra, one can define a Lie algebra by requiring that the Lie algebra product be just the commutator in the associative algebra. Do this for the Moyal product:

$$[f_1, f_2]_\tau = \sum_{a=0}^{\infty} \frac{M\tau^a}{a!} (f_1 \otimes f_2 - f_2 \otimes f_1) \qquad (5.6)$$

Of course, if τ is skew-symmetric, we see that the terms on the right hand side of (5.6) vanish for even a, hence it takes the following form:

$$[f_1, f_2]_\tau = \sum_{a=0}^{\infty} 2 \frac{M\tau^{2a+1}}{(2a+1)!} (f_1 \otimes f_2) \qquad (5.7)$$

It is then readily seen that the term on the right hand side of (5.7) corresponding to $a = 0$ is just the usual <u>Poisson</u>

bracket defined via the skew-symmetric form τ. Thus, we see that the Moyal product gives a way of "quantifying" a classical mechanical system whose symplectic structure is naturally "flat". (Notice that the group of linear automorphisms of R^n which preserve τ act as automorphisms of the Moyal product (and hence of the Poisson-Moyal Lie algebra.))

Finally, notice another virtue of this approach to quantization--its completely algebraic nature (τ can be any linear differential operator $\mathscr{F} \otimes \mathscr{F} \to \mathscr{F} \otimes \mathscr{F}$ such that formula (5.2) satisfies the associative law) --lends itself to quantization of fields. I intend to develop this approach in more detail later on.

Notice another interesting possibility inherent in the formalism--other choices of sequences of differential operators may be possible, leading to essentially different quantization schemes.

6. THE MOYAL BRACKET STANDS TO "QUANTUM SYMPLECTIC GEOMETRY" AS THE POISSON BRACKET STANDS TO "ORDINARY SYMPLECTIC GEOMETRY"

Continue with R^n and coordinates (x^i), $1 \leq i,j \leq n$. Let τ^{ij} be a constant, skew-symmetric $n \times n$ matrix which is non-singular, i.e., has nonzero determinant. (n must be even then.) Let τ_{ij} be its inverse matrix. Let

$$\omega = \tau_{ij} \, dx^i \wedge dx^j$$

Let $\mathscr{F}$ be the C^∞ real-valued functions of x. The Poisson bracket

$$\{f_1, f_2\} = \tau^{ij} \partial_i(f_1) \partial_j(f_2)$$

$$\left(\partial_i = \frac{\partial}{\partial x^i} \right)$$

then makes $\mathscr{F}$ into a Lie algebra such that

$$\{f_1, f_2 f_3\} = \{f_1, f_2\} f_3 + \{f_2, \{f_1, f_3\}\} \tag{6.2}$$

for $f_1, f_2, f_3 \in \mathscr{F}$

For $f \in \mathscr{F}$, set

$$A_f(f_1) = \{f, f_1\} .$$

(6.2) says that A_f is a derivation of $\mathscr{F}$ as a <u>commutative associative algebra</u>. The Jacobi identity says that

$$f \to A_f$$

is a Lie algebra homomorphism; also

$$df = A_f \, \lrcorner \, \omega \tag{6.3}$$

$$0 = A_f(\omega) \equiv \text{Lie derivative of } \omega \text{ by } A_f \tag{6.4}$$

Now, let $\mathscr{F}_\tau$ be <u>non-commutative associative</u> algebra defined in the following way. Its elements are <u>polynomials</u> in the variables (x^i). They satisfy (only) the following commutative relations:

$$x^i x^j - x^j x^i = \tau^{ij} . \tag{6.5}$$

Let $\mathscr{D}^r(\mathscr{F}_\tau)$ be the "differential forms" based on this algebra, whose construction is explained in previous chapters. Since the right hand side of (6.5) is a scalar, i.e., its differential is zero, the differential forms are essentially uniquely defined by the following "classical" relations:

$$dx^i x^j = x^j dx^i$$

$$dx^i \wedge dx^j = - dx^j \wedge dx^i$$

Theorem 6.1. Consider the subalgebra $P\mathscr{F}$ of $\mathscr{F}$ consisting of the polynomial functions of the x^i. Make $\mathscr{F}$ into an associative algebra using the Moyal product. Then, $P\mathscr{F}$, as an associative algebra defined via the Moyal product, is <u>isomorphic</u> to $\mathscr{F}_\tau$.

Proof. That (6.5) is satisfied in the Moyal product is obvious. Both are associative algebras composed of polynomials in the x^i, with the x^i satisfying the same commutation relations.

Thus, we can consider $\mathscr{F}_\tau$, and its differential forms. Consider

$$\omega = \tau_{ij} dx^i \wedge dx^j \in \mathscr{D}^2(\mathscr{F}_\tau) \quad .$$

It defines what we will call a <u>quantum symplectic structure</u>.

We must now extend the operations of <u>Lie derivative and inner product</u> to this "quantum" differential geometry. Here is how to do this: Let

$$A: \mathscr{F}_\tau \to \mathscr{F}_\tau$$

be a linear map which is a <u>derivation</u> of the associative (i.e., Heisenberg-Moyal) structure on $\mathscr{F}_\tau$. Thus

$$A(f_1 * f_2) = A(f_1) * f_2 + f_1 * A(f_2) \tag{6.6}$$

for $f_1, f_2 \in \mathscr{F}_\tau$,

where $*$ denotes the Moyal product. We can now extend this to <u>Lie derivative</u> and inner product operations on differential forms:

$$\begin{aligned}\mathscr{L}_A(f_1 df_2 \wedge df_3 \cdots) &= A(f_1) df_2 \wedge df_3 + f_1 dA(f_2) \wedge df_3 \wedge \cdots \\ &\quad + A \lrcorner (f_1 df_2 \wedge df_3 \cdots) \\ &= f_1 A(f_2) df_3 \wedge \cdots - f_1 df_2 A(f_3) \wedge \cdots\end{aligned}$$

Given $f \in \mathscr{F}_\tau$, we ask: Is there a derivative A_f such that

$$df = A_f \lrcorner \omega \quad ? \tag{6.7}$$

Suppose

$$f = f_i dx^i \tag{6.8}$$

Then

$$f_i dx^i = 2\tau_{ij} A(x^i) dx^j \quad ,$$

i.e.,

$$A_f(x^i) = \tfrac{1}{2} \tau^{ij} f_j \quad . \tag{6.9}$$

This formula, <u>plus</u> the condition that A act as a derivation, i.e., (6.6), now determines A completely. Then, if

$$g = a_{i_1 \cdots i_m} x^{i_1} * \cdots * x^{i_m} \quad ,$$

with

$$a_{i_1 \cdots i_m} \in R \quad ,$$

then
$$A_f(g) = a_{i_1 \cdots i_m} (A_f(x^{i_1}) * x^{i_2} \cdots * x^{i_n}$$
$$+ x^{i_1} * A_f(x^{i_2}) \cdots * x^{i_n} + \cdots$$

Given $f, g \in \mathscr{F}_\tau$, we now set
$$\{f, g\} = -2A_f(g) \quad . \tag{6.10}$$

Here is the main result:

<u>Theorem 6.2</u>. The bracket $\{\ ,\ \}_\tau$ defined by (6.10) (and intrinsically attached to the "quantum symplectic structure") is just the Poisson-Moyal bracket.

<u>Proof</u>. $A_f^\bullet$ is a derivation of $\mathscr{F}_\tau$.
$$\{x^i, f\}_\tau = 2a_f(x^i) = \tau^{ij} f_j \quad . \tag{6.11}$$

For $f, g \in \mathscr{F}_\tau$,
$$A_{f*g}(x^i) = \tau^{ij}(f*g)_j$$
$$= \tau^{ij}(f_j * g + f * g_j)$$
$$= (A_f * g + f * A_g)(x^i) \quad ,$$

i.e.,
$$A_{f*g} = A_f * g + f * A_g \quad .$$

The Moyal bracket obeys a similar rule:
$$[x^i, f*g]_\tau = x^i * f * g - f * g * x^i$$
$$= \{x^i, f\}_\tau * g + f * \{x^i, g\}_*$$

Since both Lie algebra structures agree on the linear function, they are obviously equal everywhere.

7. A NEW DEFINITION OF "QUANTUM DYNAMICS"

What is possibly most important about this material is that it gives a new way of thinking about particle quantum mechanics. Start off with an associative algebra $\mathscr{A}$. Let $\mathscr{D}^r(\mathscr{A})$ be a differential form system based on $\mathscr{A}$. Suppose that derivatives $A: \mathscr{A} \to \mathscr{A}$ of $\mathscr{A}$ can be extended to Lie-derivative type operations on $\mathscr{D}^r(\mathscr{A})$.

A <u>quantum symplectic structure</u> is now defined as a two-form $\omega \in \mathscr{D}^2(\mathscr{A})$ such that $d\omega = 0$. A <u>quantum mechanical system</u> is now defined as a derivation $A: \mathscr{A} \to \mathscr{A}$ such that

$$\mathscr{L}_A(\omega) = 0 \qquad (7.1)$$

The <u>dynamics</u> are defined by the curves $t \to a(t)$ in $\mathscr{A}$ such that

$$\frac{da}{dt} = A(a) \qquad . \qquad (7.2)$$

Bibliography

1. E.B. Elliott, _An Introduction to the Algebra of Quantics_, Chelsea Pub. Co., N.Y.
2. G. Salmon, _Lessons Introductory to the Modern Higher Algebra_, Chelsea Pub. Co., N.Y.
3. G. Whitham, _Linear and Nonlinear Waves_, J. Wiley, N.Y., 1974.

IX.

GENERALIZATIONS OF THE POISSON-MOYAL BRACKET. DIFFERENT METHODS OF QUANTIZATION

1. INTRODUCTION

The Moyal associative algebra structure and the Poisson-Moyal Lie algebra structure defined in the previous chapter obviously are tied to the <u>linear</u> symplectic structure, i.e., to R^n as a coset space

$$G/H ,$$

with G = semidirect product $Sp(n/2, R)$ on R^n, H = $Sp(n/2,R)$.

This insight immediately suggests a new setting for quantum mechanics--for example, as associated with coset spaces G/H of Lie groups which admit <u>invariant symplectic structures</u>. In this chapter I will begin to work towards such a theory.

2. GENERALIZED TRANSVECTION OPERATIONS AND MOYAL PRODUCTS

Let X be a manifold, with: $\mathscr{F}$ = the commutative, associative algebra of C^∞, real-valued functions on X. Let $\mathscr{V}$ be the Lie algebra of vector fields on X (i.e., derivations of $\mathscr{F}$). For each integer j set:

$$\mathscr{F}^j = \underbrace{\mathscr{F} \otimes \cdots \otimes \mathscr{F}}_{j \text{ times}}$$

(Continue the convention that the tensor product is <u>relative to the field of real numbers as scalars</u>.) A <u>generalized transvection operator system</u> is defined as a set

$$\tau_j : \mathcal{F} \otimes \mathcal{F} \to \mathcal{F} \otimes \mathcal{F} \quad , \qquad j = 1, 2, \ldots$$

of linear maps, such that each τ_j is a j-th order linear (i.e., multilinear on $\mathcal{F}$) differential operator.

Given such a system, one can, of course, define a Moyal structure

$$f_1 * f_2 = f_1 f_2 + \sum_{j=1}^{\infty} M\tau_j (f_1 \otimes f_2) \tag{2.1}$$

where $M: \mathcal{F} \times \mathcal{F} \to \mathcal{F}$ is the multiplication operator. In order to define a <u>quantization</u>, we must require that the *-operator defined by formula (2.1) be <u>associative</u>. Then defines a set of conditions on the operators τ_j --at this stage of things not very much is known about the possibilities.

3. THE POINT OF VIEW OF DEFORMATION THEORY

Put a real parameter $\hbar$ into (2.1):

$$f_1 * f_2 = f_1 f_2 + \sum_{j=1}^{\infty} \hbar^j M\tau_j (f_1, f_2) \quad .$$

Set:

$$\delta(f_1, f_2) = \frac{\partial}{\partial \hbar} (f_1 * f_2) \big|_{\hbar = 0}$$

$$\delta_2(f_1,f_2) = \frac{\partial^2}{\partial \hbar^2}(f_1*f_2)\Big|_{\hbar=0}$$

and so forth. Hence,

$$\delta(f_1,f_2) = M\tau_1(f_1,f_2)$$

Differentiate the associativity relation to the second order:

$$(f_1*f_2)*f_3 = f_1*(f_2*f_3)$$

$$\delta(f_1*f_2,f_3) + \delta(f_1,f_2)*f_3 = \delta(f_1,f_2*f_3) + f_1*\delta(f_2*f_3) \quad (3.1)$$

Now set $\hbar = 0$, obtaining the following <u>first order conditions</u>:

$$\delta(f_1 f_2, f_3) + \delta(f_1,f_2)f_3 = \delta(f_1,f_2 f_3) + f_1 \delta(f_2,f_3) \quad (3.2)$$

Differentiate (3.1) with respect to $\hbar$, then set $\hbar = 0$:

$$\delta_2(f_1 f_2, f_3) + 2\delta(\delta(f_1,f_2),f_3) + \delta_2(f_1,f_2)f_3$$
$$= \delta_2(f_1, f_2 f_3) + 2\delta(f_1,\delta(f_2,f_3)) + f_1 \delta_2(f_2 f_3) \quad (3.3)$$

Since

$$\delta(f_1,f_2) = M\tau(f_1 \otimes f_2)$$
$$\delta_2(f_1,f_2) = 2M\tau_2(f_1 \otimes f_2) \,, \quad (3.4)$$

we see that relations (3.2), (3.3) (and those obtained by further differentiation) gives a set of conditions on the τ_j. (These have a well-known cohomological interpretation, but I do not see, at least at the moment, that this particularly helps.)

Let us now analyze condition (3.2). Our assumption about the τ's (as differential operators) implies that δ satisfies the following conditions:

$$\delta(f_1 f_2, f_3) = (f_1 \delta(f_2, f_3)) + f_2 \delta(f_1, f_3)$$
$$\delta(f_1, f_2 f_3) = \delta(f_1, f_2) f_3 + \delta(f_1, f_3) f_2$$
(3.4)

Substitute relations (3.4) into (3.2), obtaining:

$$f_2 \delta(f_1, f_3) = \delta(f_1, f_3) f_2 \quad ,$$

which is satisfied identically. We can sum up as follows:

<u>Theorem 3.1</u>. The first order conditions for associativity of the *-product, namely condition (3.2), are a consequence of the following assumption about δ:

> For fixed f_1, f_2, the map $f \rightarrow \delta(f, f_2 \delta_2)$, $f \rightarrow \delta(f_1, f)$ are derivations of $\mathscr{F}$ into itself, i.e., first order linear differential operators.
(3.5)

Condition (3.5) can also be interpreted "geometrically":

<u>Theorem 3.2</u>. Let $\mathscr{F}^1$ denote the one-differential forms on X, considered as an $\mathscr{F}$-module. Condition (3.5) means that there is an $\mathscr{F}$-bilinear map $\omega: \mathscr{F}^1 \times \mathscr{F}^1 \rightarrow \mathscr{F}$ such that:

$$\delta(f_1,f_2) = \omega(df_1,df_2) . \tag{3.6}$$

ω defines a tensor-field, i.e., a "geometric structure" on X.

Let us turn to the analysis of the second order condition (3.3). Our assumption about τ_2 means that δ_2 takes the following form:

$$\delta_2(f_1,f_2) = \sum a^{ij} D_i(f_1) D_j(f_2) , \tag{3.7}$$

where D_i are second-order linear differential operators: $\mathscr{F} \to \mathscr{F}$, a^{ij} are constants. They satisfy the following general relation:

$$D_i(f_1 f_2) = D_i(f_1) f_2 + D_i(f_2) d_1 + \Delta_i(f_1,f_2) ,$$

where $\Delta_i : \mathscr{F} \times \mathscr{F} \to \mathscr{F}$ is a first order, <u>bilinear</u> differential operator. Thus, we have

$$\delta_2(f_1 f_2, f_3) = \sum a^{ij}(D_i(f_1) f_2 + f_1 D_i(f_2) + \Delta_i(f_1,f_2)) D_j f_3$$

$$= \delta_2(f_1,f_3) f_2 + f_1 \delta(f_2,f_3) + \sum a^{ij} \Delta_i(f_1,f_2) D_j(f_3)$$

Set:

$$\Delta(f_1,f_2,f_3) = \delta_2(f_1 f_2,f_3) - \delta_2(f_1,f_3) f_2 - f_1 \delta_2(f_2,f_3)$$

We see that Δ is an R-linear map

$$\mathscr{F} \times \mathscr{F} \times \mathscr{F} \to \mathscr{F} ,$$

which is first order in f_1 and f_2, second order in f_3.

Now, in order to simplify the analysis, let us assume that the * operator has the same symmetry features as the traditional Moyal product, i.e.,

$$\delta(f_1, f_2) = -\delta(f_2, f_1) \tag{3.9}$$

$$\delta_2(f_1, f_2) = \delta_2(f_2, f_1) \quad . \tag{3.10}$$

Condition (3.9) means that the tensor field ω is skew-symmetric. Let us further assume that the geometric structure defined by ω is "integrable", i.e., that the product

$$(f_1, f_2) \to \delta(f_1, f_2) = \omega(df_1, df_2)$$

satisfies Jacobi identity, i.e., defines a <u>cosymplectic structure</u> on X. (See "Interdisciplinary Mathematics", Volumes 10 and 15.)

<u>Remark</u>. It is interesting that there is a general geometric reason for calling this the "integrability condition". ω defines what differential geometers call a "G-structure" on M. This defines a "reduction of the structure group of the principal tangent bundle to M". (Actually, this is a more general object--the structure group may vary from point to point on X. The <u>general</u> differential geometric machinery has not been worked out for this case. It would, in fact, be very useful for many physical and geometric situations. I plan to work on this at some point.)

Return now to the second order condition (3.3). The Jacobi identity, i.e., the integrability condition for δ, is the following:

$$\delta(f_1,\delta(f_2,f_3)) = \delta(\delta(f_1,f_2),f_3) + \delta(f_2,\delta(f_1,f_3)) \quad (3.11)$$

Substitute this into (3.3), obtaining the following identity:

$$\delta_2(f_1f_2,f_3) + \delta_2(f_1,f_2)f_3 = \delta_2(f_1,f_2f_3) + 2\delta(f_2,\delta(f_1,f_3)) + f_1\delta_2(f_2,f_3) \quad (3.12)$$

Now, substitute (3.8) into (3.12), and use the Ansatz that δ_2 is symmetric:

$$\Delta(f_1,f_2,f_3) + \delta_2(f_1,f_3)f_2 + f_1\delta_2(f_2,f_3) + \delta_2(f_1,f_2)f_3$$

$$= \Delta(f_2,f_3,f_1) + \delta_2(f_2,f_1)f_3 + \delta_2(f_3,f_1)f_2$$

$$+ 2\delta(f_2,\delta(f_1,f_3)) + f_1\delta_2(f_2,f_3)$$

Luckily, there are extensive cancellations, and we are left with the following relation:

$$\boxed{\Delta(f_1,f_2,f_3) - \Delta(f_2,f_3,f_1) = 2\delta(f_2,\delta(f_1,f_3))} \quad (3.13)$$

This is the definitive condition (with the various reasonable simplifying assymptions we have made) that the $*$ product be associative up to <u>second order in</u> $\hbar$.

How successfully we could continue in this way to definitively and _usefully_ analyze the conditions that interesting *-products of this type exist is not clear to me at this stage. Perhaps a necessary preliminary would be the exhibition of more interesting examples. No doubt these will be produced most readily using Lie group-theoretic considerations. I now turn to such examples.

4. THE MOYAL STRUCTURE ON THE DUAL SPACE OF A LIE ALGEBRA

Let $\mathscr{G}$ be a (finite dimensional) Lie algebra (over the real numbers as scalars, say). Let

$$X = \mathscr{G}^d \equiv \text{dual space of } \mathscr{G} \quad .$$

We know (Interdisciplinary Mathematics, Volumes 10 and 15) that X has a natural cosymplectic structure. We can then attempt to use it to define a Moyal structure.

Here is how it is defined. Let A^i, $1 \leq i,j \leq n$, be a basis for $\mathscr{G}$, with structure relations:

$$[A^i, A^j] = C^{ij}_k A^k \quad . \tag{4.1}$$

Let x^i be the corresponding linear function on $\mathscr{G}^d$ (having the usual isomorphism $(\mathscr{G}^d)^d \approx \mathscr{G}$). If the cosymplectic structure is denoted by $\{\ ,\ \}$, then

$$\{x^i, x^j\} = C^{ij}_k x^k \quad . \tag{4.2}$$

Let us now try to construct a Moyal structure. Set:

$$\tau(x^i \otimes x^j) = \frac{1}{2} C_k^{ij}(x^k \otimes 1 + 1 \otimes x^k) \qquad (4.3)$$

Let $\mathscr{F}$ denote the commutative associative algebra consisting of the polynomials on the x^i. Requiring that τ be a first order bilinear differential operator then defines it as an R-bilinear map

$$\tau: \mathscr{F} \otimes \mathscr{F} \to \mathscr{F} \otimes \mathscr{F} \ .$$

Suppose

$$A = a_i x^i \ ; \qquad B = b_i x^i$$

are elements of $\mathscr{G}$. (Continue to identify $(\mathscr{G}^d)^d$ with $\mathscr{G}$.) Then (4.3) takes the following coordinate-free form

$$\tau(A \otimes B) = \frac{1}{2} ([A,B] \otimes 1 + 1 \otimes [A,B]) \qquad (4.4)$$

We can now reformulate the construction in a more coordinate-free and <u>algebraic</u> way.

Let $\mathscr{G}$ be a real Lie algebra (not necessarily finite dimensional any longer). Let

$$S(\mathscr{G})$$

be the commutative, associative algebra formed by the <u>symmetric</u> tensors on $\mathscr{G}$. (The algebraic product is then the symmetric tensor product; see IM, Vol. 2.) Define an R-linear mapping

$$\tau: S(\mathscr{G}) \otimes S(\mathscr{G}) \to S(\mathscr{G}) \otimes S(\mathscr{G}) \qquad (4.5)$$

as follows:

On the subspace $\mathscr{G} \otimes \mathscr{G}$ of $S(\mathscr{G}) \otimes S(\mathscr{G})$ (i.e., the terms of gradation one), τ is defined by (4.4). It is extended to all of $S(\mathscr{G}) \otimes S(\mathscr{G})$ by the condition that it be a bilinear, first order differential operator, i.e., for one argument held fixed, it is a derivation when the other is varied.

This τ-operator is a <u>generalization of the transvection operator of classical invariant theory</u>. It reduces to it if $\mathscr{G}$ is taken as the Heisenberg Lie algebra. We can then form the "Moyal structure" by analogy with what was in the case $\mathscr{G}$ = Heisenberg Lie algebra

$$f_1 * f_2 = \sum_{j=0}^{\infty} M\tau^j (f_1 \otimes f_2) \quad , \tag{4.6}$$

where τ^j is the j-th power of τ, $f_1, f_2 \in S(\mathscr{G})$, and $M: S(\mathscr{G}) \otimes S(\mathscr{G}) \to S(\mathscr{G})$ is the multiplication operation, i.e.,

$$M(f_1 \times f_2) = f_1 f_2 \ (\equiv f_1 \circ f_2 \equiv \underline{\text{symmetric}} \text{ tensor product!})$$

We can only compute the $*$ operator explicitly for the first order term: For $A, B \in \mathscr{G}$,

$$\tau^2 (A \otimes B) = \tau(\tau(A \otimes B))$$

$$= \tfrac{1}{2} \tau([A,B] \otimes 1 + 1 \otimes [A,B])$$

$$\equiv 0 \quad ,$$

since τ is a <u>differential operator</u>, i.e., the derivative of "1" is zero. Hence,

$$A*B = AB + [A,B] \tag{4.7}$$

We can now exhibit a reasonable <u>candidate</u> for the Moyal structure. Let

$$\rho: \mathscr{G} \to L(V)$$

be a linear representation of $\mathscr{G}$ by linear maps on a vector space V. Define a map

$$\rho: S(\mathscr{G}) \to L(V)$$

as follows:

$$\rho(A_1 \cdots A_m) = \frac{1}{m!} \sum_{\substack{\text{(all permutations} \\ \text{of } (1,\ldots,m)\,)}} \rho(A_{i_1}) \cdots \rho(A_{i_m}) \tag{4.8}$$

<u>Remark</u>. In order to make the formulas balance out, we must define the "representation of $\mathscr{G}$" condition for ρ to be the following identity:

$$\rho([A,B]) = \frac{1}{2}(\rho(A)\rho(B) - \rho(B)\rho(A)) \tag{4.9}$$

With the convention (4.9) in force, we can work out:

$$\rho(A*B) = \frac{1}{2}(\rho(A)\rho(B) + \rho(B)\rho(A)) + \frac{1}{2}(\rho(A)\rho(B) - \rho(B)\rho(A))$$

$$= \rho(A)\rho(B) \quad . \tag{4.10}$$

(4.10), of course, suggests the following statment:

> ρ, as defined by (4.8), is a homomorphism of the *-algebra structure on $S(\mathscr{G})$ into the associative algebra $L(V)$. In particular, * is itself associative.

(4.11)

My private calculations convince me that (4.11) is true, but I am not a good enough combinatorist to complete the proof. At any rate, I would be very surprised if it were false! Further work will be provided in later volumes.